# Health Informatics

This series is directed to healthcare professionals leading the transformation of healthcare by using information and knowledge. For over 20 years, Health Informatics has offered a broad range of titles: some address specific professions such as nursing, medicine, and health administration; others cover special areas of practice such as trauma and radiology; still other books in the series focus on interdisciplinary issues, such as the computer based patient record, electronic health records, and networked healthcare systems. Editors and authors, eminent experts in their fields, offer their accounts of innovations in health informatics. Increasingly, these accounts go beyond hardware and software to address the role of information in influencing the transformation of healthcare delivery systems around the world. The series also increasingly focuses on the users of the information and systems: the organizational, behavioral, and societal changes that accompany the diffusion of information technology in health services environments.

Developments in healthcare delivery are constant; in recent years, bioinformatics has emerged as a new field in health informatics to support emerging and ongoing developments in molecular biology. At the same time, further evolution of the field of health informatics is reflected in the introduction of concepts at the macro or health systems delivery level with major national initiatives related to electronic health records (EHR), data standards, and public health informatics.

These changes will continue to shape health services in the twenty-first century. By making full and creative use of the technology to tame data and to transform information, Health Informatics will foster the development and use of new knowledge in healthcare.

More information about this series at http://www.springer.com/series/1114

Egondu R. Onyejekwe • Jon Rokne
Cory L. Hall

Editors

# Portable Health Records in a Mobile Society

 Springer

*Editors*
Egondu R. Onyejekwe
The School of Health Sciences
Walden University
Minneapolis, MN
USA

Jon Rokne
Department of Computer Science
University of Calgary
Calgary, AB
Canada

Cory L. Hall
University of Miami
Pembroke Pines, FL
USA

ISSN 1431-1917                ISSN 2197-3741   (electronic)
Health Informatics
ISBN 978-3-030-19936-4        ISBN 978-3-030-19937-1   (eBook)
https://doi.org/10.1007/978-3-030-19937-1

This Springer imprint is published by the registered company Springer Nature Switzerland AG
The registered company address is: Gewerbestrasse 11, 6330 Cham, Switzerland

# Preface

Medical records are some of the most complex records ever created. The content in a medical record ranges from quick notes about a patient's activities of daily living to the nuclear physics of proton therapy and the analysis of genetic sequences. They can range from just a single page to thousands of pages and electronically stored information too voluminous and detailed to ever be presented in physical form. All of the information in a specific medical record is focused on a specific individual that has limited access to, or control over, the content and how that content is managed. This book explores topics related to patient-directed access and management of their medical records. No single book could address the topic comprehensively. Instead, the topics in this book provide a broad spectrum of information that can inform the interested and serve as a jumping-off point for further investigation.

Historically, a medical record was stored as an organized physical document. Many years ago, when I was in the Air Force, these physical documents were sealed in a large envelope and given to me to deliver, unopened, to the medical service at my newly assigned base. This was my first and only experience with a truly portable medical record. Although I had no control over the content of that record and would have disobeyed orders if I looked inside, for a short period of time, I had personal possession of my entire medical record. Even X-rays were stored as physical films until the use of tomographic analysis became ubiquitous. (Fun fact, a source of revenue for hospitals in the past was silver recovery from discarded X-ray films.) The advent of tomography signaled the beginning of the end of the physical medical record. Multidimensional imaging cannot be effectively stored and used in a physical form. From that point forward, the medical record has become more of a convenient mental model than a reality.

Today's medical record is mostly electronic. Even physical documents are scanned and stored electronically rather than physically. Medical records departments in modern hospitals have been renamed as health information management, or as a newer term health information integrity, department to reflect the changing role of that department from physical document management to stewardship of the electronic records of the patients they serve.

At the same time, ownership of the record has transitioned from mostly care provider owned to mostly patient owned. For years, the record was created by care providers for use by care providers and was fiercely defended as such by provider-supported associations. As the source of documentation became more diversified and distributed, the argument for focused ownership began to lose ground. The advocacy for patient rights has made significant progress with more and more rights being given to patients or their designated representatives.

Today, patients are gaining more control over their care and accompanying documentation. With increased control comes increased responsibility. How do we support these new rights? What does the patient need to know about their medical record? What tools are needed to support the patient? What are the practical steps we need to take? What obstacles stand in the way? How do we overcome those obstacles? This book seeks to explore those questions and possible answers.

The other editors and I would like to thank the many authors and contributors for the many hours spent writing, editing, and formatting the content. We dedicate this book to them and the many people seeking to empower patients by giving hours and hours of their time in this cause.

Here, we provide special recognitions to some people. While we acknowledge all of the contributors toward this effort, the following people and/or entities, in particular, helped make this book possible:

Dr. Lawrence Okey Onyejekwe Jr. —whose original thoughts in a start-up, Zobreus.com, stimulated and inspired the subject matter and, indeed, the title of the book

Springer—who awarded the contract and provided the approval to go on with the subject matter and topic; they also moved the project by providing additional clerical support at their end

Dr. Hung Ching—a contributor, who also helped with the editing and formatting of the manuscript and who became the custodial of the chapters as they progressed toward completion

Dr. Dasantila Sherifi—another contributor, who also assisted with editing and typesetting and who took over as the final custodian when Dr. Ching was away

We hope you will enjoy reading this book!

Pembroke Pines, FL, USA                                              Cory L. Hall

# Contents

**Part VII   Future Directions**

**Part VIII   Conclusion**

# About the Editors

**Egondu R. Onyejekwe** is a professor of Health Informatics who holds six degrees in a variety of fields, starting with a BSc (Hons) in Zoology (Parasitology), University of Nigeria, Nsukka. She holds three additional masters' degrees from The Ohio State University (Columbus, Ohio) in Zoology (MS), Health Education (MA), and Public Health Administration (MA). From Ohio University, Athens, Ohio, she earned an MS degree in Industrial and Systems Engineering (Software Engineering) and a doctorate degree in Information Systems (Communications). She has served in different capacities and in different institutions, including The Ohio State University, the World Health Organization, among others. She served as a visiting professor of Health Informatics at the University of Technology Services in Nigeria and retired from The Ohio State University, Columbus Ohio. She is currently a part-time faculty at Walden University. She serves as the overseas liaison officer for Gregory University, Uturu (GUU), Abia State, Nigeria. She is also an associate editor, Springer Publishing, who commissioned her to serve as the editor of this book on Portable Health Records. She is certified by the AHIMA (American Health Information Management Association) as a HIT Pro trainer (health information technology trainer). Her research interests focus on big data and big data analytics in healthcare.

**Jon Rokne** is a professor in the Department of Computer Science at the University of Calgary where he served as the department chair from 1989 to 1996. His research interests include interval analysis, global optimization, computer graphics, and social networks. He is a coeditor of the prestigious *Encyclopedia of Social Network Analysis and Mining* (2014 with second edition in 2018). He organized the installation of the first university FDDI network in Canada at the University of Calgary, and in 1990, he actively worked toward connecting the university to the World Wide Web (WWW). He spearheaded the acquisition and installation of a top 500 supercomputing facility in 1992. An IEEE member since 1970 and life fellow since 2013, he has volunteered for IEEE Computer Society where he completed two terms as vice president on Publications and three terms on the Board of Governors. He has also served as vice president, IEEE Publication Services and Products Board, and as a member of the Board of Directors of the IEEE for two terms.

**Cory L. Hall** is an assistant vice president of Information Technology for the University of Miami Health System. He has over 30 years of experience working for healthcare provider organizations and healthcare information technology vendors. He has helped design, build, implement, and support almost every type of operational IT system in common use at healthcare organizations around the world. He has worked as a consultant for Cap Gemini Ernst and Young, managed the clinical products at Eclipsys (now Allscripts), built the world's first health information exchange for Ameritech Health Connections, served as the director of Diagnostic Intelligence for the College of American Pathologists, and managed diagnostic imaging products at Merge Healthcare (now IBM) and several venture capital-funded start-ups focused on innovative healthcare applications like telemedicine. Although he now calls Miami, Florida, home, he has lived and worked in many locations across the United States, Canada, Europe, the Middle East, and India.

# About the Authors

**Lovette Chinwah-Adegbola, Ph.D.** is a full professor of Journalism and Mass Communication and director of the Journalism and Digital Media Program at Central State University, Wilberforce, Ohio. She has served as a faculty fellow with numerous professional organizations. She earned her B.S. and M.A. degrees from the University of Wisconsin-Superior and her Ph.D. from the Ohio University and subsequently completed the Harvard University Management Development Program.

**Cheryl Austein Casnoff, MPH** is a senior fellow at NORC at the University of Chicago. She has had a distinguished public service career in public health, health financing, and health IT and a deep understanding of federal, state, local, and tribal government public health and health systems policies. The highlights of her career include establishing and leading a new Office of Health Information Technology (OHIT) within the Health Resources and Services Administration (HRSA) to promote the adoption and meaningful use of electronic health records for safety-net providers. She has a BS in Biological Sciences from Northwestern University and an MPH from Yale School of Medicine Department of Epidemiology and Public Health.

**Hung Ching** is a board-certified and licensed senior medical physicist at Memorial Sloan Kettering Cancer Center in New York City. He received his Bachelor of Science and Master of Science degrees in physics from the State University of New York at Stony Brook. Moreover, he also has a Doctor of Philosophy degree in Public Health from Walden University.

**LaQuasha Gaddis** graduated in 2001 with a Bachelor of Arts degree in Biology from the University of California, Riverside. After moving to Georgia, she worked at Georgia State University for 3 years in the Biology Department and had the opportunity to be a coauthor on three scientific journal publications. In 2005, she transitioned to work as a chemist on environmental health issues, specifically studying exposure to second-hand smoke, in the Division of Laboratory Sciences for the

Centers for Disease Control and Prevention (CDC). She currently works at the CDC as a Public Health Analyst for the Division of Foodborne, Waterborne, and Environmental Diseases.

**Jefferson Howe** is a consultant in Health Information Management with more than 35 years in the healthcare industry. He holds a master of science degree with concentration in organizational behavior from St. Michael's College in Colchester, Vermont, and is a registered health information administrator.

**Solomon Koppoe** received his undergraduate degree in Business Administration, Finance, from the University of Massachusetts, Amherst, and received a Master of Science degree in Technology Management with Federal CIO Certification from the George Mason University Graduate School of Management. He completed his Doctorate in Healthcare Administration in March 2018 at Walden University. He spent about 10 years in the telecom industry as a senior manager prior to pursuing his graduate degree in Technology Management. He currently works as a consultant in IT Project Management.

**Dorcas Maina** is a clinical nurse who lives in Kenya and has been teaching in nursing for the last 15 years. Her focus is in emergency care, critical care, and trauma nursing with an interest in access to care. Furthermore, she has an interest in exploring the use of technology among this and other populations to address health equity. In this capacity, she is focusing on technology use among patients with chronic illness in her dissertation.

**Allison Nnaka** is the director of Nursing at the IAFF Recovery Center. Her nursing clinical competency spans from emergency care, rehabilitation care, and adult medical/surgical care. In addition to her 20 years of clinical experience, she also has an outstanding academic background that includes a bachelor's degree in Nursing and a master's degree in Nursing and in Health Administration from the University of Phoenix. She is currently a PhD candidate at Walden University.

**Dasantila Sherifi** is a professor of Health Information Management and Health Services at DeVry University and a registered health information administrator. She holds a bachelor's degree in Marketing and in Health Information Administration from the University of Tirana and Gwynedd Mercy University, Pennsylvania, respectively; a masters' degree in Business Administration from Southern Illinois University Carbondale, Illinois; and a Doctor of Philosophy degree in Health Services from Walden University.

# Part I
# Introduction

# Chapter 1
# Introduction: The Context of Time and Space

Egondu R. Onyejekwe

**Abstract**  This book begins by adopting the Western Hemisphere's concept of *time* as being composed of the *past*, *present* and the *future* as well as being tied to the concept of *space*. The concept of *space*, in physics, has three dimensions, which as a spacetime, becomes a mathematical model that fuses the three space dimensions of length, width and height, with the one dimension of time, yielding a single four-dimensional continuum. The acceptance of the process that involves time and space as defined in everyday usage, still creates an additional layer called time and space complexity. So, regardless of how time and space are defined, they are not easy concepts to grasp. A medical or a health record, sits in some defined space, and is processed as it is ported and or delivered from the beginning of some time to the end of a defined time. It would include space complexity involving the amount of computer *memory*—fixed and variable memory—required to solve the problem of any given *size* as a result of that porting. It would also depend on the running time or time complexity as measured by the size of the input data, the hardware, the operating system and the programming language that is used. The running time would yield some cost. When and where the records are varied and fragmented, the costs escalate. Therefore, regardless of how time and space are perceived, conceived, or defined, it is clear that a *portable* health record in a *mobile society* is a complex matter. Today's electronic health record system maintains an isolated sliver of a patient's health record. But, today's patient as a healthcare consumer, wants healthcare commoditized and expects similar levels of convenience, consistency, personalization and reliability as provided by other consumer services. There is an additional need to capture the health records carried by patients and include them in the electronic record system. The need is to move the formal and fragmented record from the confines of its local EHR to a common, secure, distributed and accessible space where afterwards (time is expensed) it becomes useable. An emerging technology, the Blockchain distributed ledger technology used by BitCoin seems to address that problem. Artifical Intelligence (AI) using machine learning algorithms provides a way to extract salient health record data. Also, the Blockchain technology addresses

E. R. Onyejekwe (✉)
College of Health Sciences, Walden University, Minneapolis, MN, USA
e-mail: Egondu.onyejekwe@mail.waldenu.edu

© Springer Nature Switzerland AG 2019
E. R. Onyejekwe et al. (eds.), *Portable Health Records in a Mobile Society*,
Health Informatics, https://doi.org/10.1007/978-3-030-19937-1_1

3

interoperability challenges, is based on open standards, and provides a shared distributed view of health data that remains private and secured! Towards that end, Electronic Health Record systems could be transformed into systems that access and update a common patient specific branch of a healthcare focused distributed ledger. Each entry in that health ledger would then be validated by the contributor, secured by the platform and authorized by the patient or their agent. The Blockchain technology would engage millions of individuals, health care providers, health care entities and medical researchers who would also like to share vast amounts of a variety of data. There is thus, an urgent need and it seems imperative to commoditize electronic health record! Where this aspect of healthcare becomes a commodity, that would imply a uniform and universally accepted definition of the end product. Maybe, that would curtail and or offset the spiraling cost of the US Healthcare system, and it can be done within the confines of spacetime!

**Keywords** Time · Space · Spacetime · Universe · Blockchain · Artificial Intelligence (AI) · Machine learning algorithm · Bitcoin · Commoditization Electronic health records · Electronic medical records

## 1.1   Introduction

"The only way to discover the limits of the possible is to go beyond them into the impossible (Clarke 2018)." Hopefully, that's what we are doing with the electronic health record at this time. This book begins by adopting the Western Hemisphere's concept of *time* as being composed of the *past, present* and the *future* as well as being tied to the concept of *space*. If the *past*, in some senses, is such as that of a distant star that sees our present, then it must exist and is not totally lost. Furthermore, if the past could be changed, then in this posit, that past would not be the past of our current *present*. So, if we hold that the *past* can be changed, that would have implications for our *present*. This brings up the concept of the *present*, which we explain as the point of flow of the *present* into the *past*. But is the present quantifiable? A snapshot provides a picture of which the *present* is the scene itself composed however, of small time differences of the different elements and parts of the image on the scene. Each of these captures a future that disappears into the *past*. If so, then, the present is not exactly defined in the scene, as it keeps encroaching on the future. Does that make the *future* static? From the image capture of a scene, we know already that we can change the *future* by changing the *present*. (The assumption here is that we know what the present is.) The implications of the change, however defy our current senses and explanation models, because we cannot explain them. Bring in the concept of *space*, which in physics has three dimensions, of distance that are measured as: length, width, and height. A spacetime, however, is a mathematical model that fuses the three dimensions of length, width and height, with the one dimension of time, yielding a single four-dimensional continuum (SpaceTime 2017). There are as varied opinions as there are experts about the existence of such a four-dimensional entity. Even with the acceptance of time and space as defined in everyday usage, the processes that involve time and space, create an additional layer called time and space complexity (http://lets-learncs.com/time-and-space-complexity/). This is because every real-world process depends on how much time as calculated—in the arrow from past, present to future—it would take to execute as well as how much space the process consumes.

It is against this backdrop that we explore the concept of a portable health record in a mobile society. Today each electronic health record system maintains an isolated sliver of a patient's health record. What if that formal and fragmented record is moved from the confines of the local EHR to a common, secure, distributed and accessible health record? On a conceptual level, such a common, secure, distributed and accessible platform could be implemented using the Blockchain—a peer-to-peer (P2P) distributed ledger technology used by BitCoin and one that is useful to healthcare because this new generation of transactional applications establishes both transparency and trust. As the underlying fabric for Bitcoin, Blockchain is a design pattern of three main components: a distributed network; a shared ledger; and digital transactions. Electronic Health Record systems could be transformed into systems that access and update a common patient specific branch of a healthcare focused distributed ledger. Each entry in that health ledger would then be validated by the contributor, secured by the platform and authorized by the patient or their agent. Because the Blockchain technology addresses interoperability challenges, is based on open standards, and provides a shared distributed view of health data that remains private and secured, it is likely to encourage widespread acceptance. This would engage millions of individuals, health care providers, health care entities and medical researchers who would also like to share vast amounts of genetic, diet, lifestyle, environmental and health data with guaranteed security and privacy protection. Blockchain technology definitely has a place in the US healthcare IT ecosystem. Hopefully, the Office of the National Coordinator (ONC) that has spurned the unwieldy growth of EHR by incentivizing it should strongly consider basing their interoperability strategy on Blockchain and using Blockchain to promote the advancement of all of healthcare.

Health economist decry the cost of Healthcare in the U.S. for several reasons, among which is its fragmentation. This is not noticeable to the naked eye, because fragmentation is a by-product of segmentation and differentiation in the healthcare marketplace. Each provider through a fragmented part of the market may have well-intentioned actions that sometimes have the unintended consequence of making things worse for the whole landscape of care (May 2001). Such unintended consequences include: inefficiency; ineffectiveness; inequality; escalated costs among others (Stange 2009). To address these consequences, the U.S. Healthcare ecosystem, should consider *commoditization*. This would begin with Blockchain technology that runs on widely used and reliable commodity hardware. Commoditized hardware would provide the greatest amount of useful computation at low cost. The hardware would be based on open standards and would be manufactured by multiple vendors. This would be the most cost effective and efficient architecture for health, healthcare and other areas research. Additionally, excess Blockchain hardware capacity could be shared with diverse field of health researchers and would facilitate faster discovery of new drugs and treatments. Furthermore, Electronic Health Records (EHRs) need to be commoditized—that is as goods coming from different providers, they should be so similar in design and function that they are set apart in price only by slight slight variations. Obviously, today's healthcare IT is not currently a commodity in the sense that many IT devices and services are. So, the hospitals operate without information regarding the standard components of a good EHR. Neither do they have ideas of what reasonable prices look like, nor what functions are truly necessary and what functions are not needed. Finally, it is driven by financial incentives and group think rather than complete information. So, this market is classified as imperfect in economic terms! So, and in oversimplified way, healthcare IT must realize agreed-upon modularity and standards, which should lead to commoditization in technology and experts agree that the process of commoditization can still have an impact on healthcare. For the consumer, though, commoditization is good when it drives down prices, but when it seems to stifle innovation, it may not be so good.

## 1.2   Time

In a personal communication with Jon Rokne, as he was attending the IEEE Technology Time Machine (TTM) 2016, he stated, "At the TTH, I was thinking a bit more about time. There are three aspect of time: past, present and future. "Past, we cannot change. Present, not clear how it fits. Future, can be changed, but not predictably [Rokne J 2016, personal communication, 22nd November].' " (By the way, for anyone interested in the technology of time machine, the IEEE Future Directions Committee [FDC], which is an incubator for emerging technologies, will be organizing its fifth symposium on the IEEE Technology Time Machine [TTM] in 2018. Also, the IEEE Technology Time Machine [TTM]  2016, had interesting sessions that focused on big data, cybersecurity, the Internet of Things, the cloud, the brain, and rebooting computing, as well as a presentation on women's roles in making the future.)

I have been fascinated since childhood by the concept of time. This is not only because time presents to us in many different ways, but also because time may not really have any existence! This is a riddle I have embraced for decades. Born a Nigerian, the way my family and the rest of the country viewed, time was a far cry from the Western orientation I learned from school. My culture perceives time as being bi-directional. We have a 4-day week that recycles with the rising and the setting of the sun. But we also have reincarnation, which ascribes life of a new-born to that of someone who already lived in the past! So, everyone born today, already lived a life in the past!

Enter my physics classes/courses! There, for example, I learned that time plays a major role in the measurement of motion and forces. We were taught that a major breakthrough in the understanding of time occurred about one hundred years ago with Einstein's theories of special and general relativity. Einstein's general theory of relativity holds that about 13.7 billion years ago (Science daily: what is time? 2005), an event called the Big Bang occurred. Space, or the universe, emerged in the Big Bang at that point. Prior to the Big Bang, matter as we know it, was simply packed into a very tiny dot (Science daily: what is time? 2005). From that dot, also emerged the part of matter that later became the Sun, the Earth, and the Moon. These, of course, are the heavenly bodies that tell us about the passage of time. Time itself barely has an independent existence, but is manifested through change. Such a change would include the circular motion of the Moon around Earth. The passage of time is, indeed, closely connected to the concept of space. Therefore, time passes ceaselessly, and we can follow it with clocks and calendars (Wikipedia: time in physics 2018). But the tools we have (such as microscope[s]) are not adequate to study time, and consequently, we cannot experiment with time only because it keeps passing. Worse yet, it is not feasible to document exactly what happens when time passes (Science daily: what is time? 2005)!

Howbeit, Einstein's theories introduced the concept of how time slows with both motion and in gravity. Furthermore, Einstein also showed that "large masses curve space and there is increase in mass with acceleration by application of a force (Tick-Tock 2017)." We surmise that the riddle of time can then be solved by these discoveries of Einstein (Christoforou 2014). But wait—Not every culture perceives this view of time.

For example, while the Western cultures, especially the US and most of Western Europe embrace Einstein's concept of time, not every culture or country perceives time that way. The western cultures also have a linear perception of time that delineates three aspects of time as: past, present and future. Therefore, time progresses from the past, into the present and into the future.

However, this culture of time does not permit us to change the past (Howell 2018). Yet, we are very unclear about how we can fit the present. At the same time, while the future can be changed, this is not in predictable ways. So, here's what we get:

***Past*** What is the past? Is it lost? What about a star far away? It 'sees" our present. So, maybe the past does exist in certain senses. Also, if the past could be changed it would not be the past of our present. So, it can be changed, possibly, but this has implications about the present. So, then we move to the present.

***Present*** It is the point of flow of present into past. But how can this be quantified. If I take a picture of a scene I see the picture, the camera sees the picture and the present is the scene itself. There are, however, small time differences between all of the items resulting in the image. This means that I do not know exactly what is defined as the present. Was the future captured as well?

*Future* This is not static. We can change it by changing the present assuming we know what the present is, but we do not know what the implications of the change are.

Rather than be caught up in this mesh of uncountable units: the Western culture simply tries to "measure" time by instantiations and schedule events with *near* precisions such as the flight schedule, train services, bus services and even meetings, thereby wrapping their daily lives around the "onward rush of time." To these cultures, then, the many facets of time suggest that time is an emergent phenomenon: in other words, that time arises from some underlying process that needs to be identified.

Yet, time is seen in a particularly different light by Eastern and Western cultures and even within the same blocks one encounter dissimilar aspects of time from country to country. For example, in the Western Hemisphere, the United States and Mexico employ time in diametrically opposing ways, which causes friction especially in businesses between these two cultures. The same is true in Western Europe, where the Swiss attitude of time conflicts with that of neighboring Italy. Furthermore, Thais do not evaluate the passing of time as the Japanese do. At another extreme, is that in Britain the future stretches out in front of you, whereas in Madagascar it flows from behind into the back of your head!

Still deeper down and for some cultures, such as people of the Piraha heritage in the Amazon, there is no concept of time beyond the present. To them, the future couldn't possibly exist. Since time is also tied to language, they do not even have a word for the concept of "future" in their vocabulary! They just live for the present!

Furthermore, the average human being (except of course, for the Piraha of the Amazon) finds it impossible to picture any other extraterrestrial culture, where for instance, the arrow of time—that is, the direction of time flows from past to future—is reversed. This would imply that time as we know it, flows from the future into the past. While this is hard to envision, that doesn't mean that it cannot be, or that time cannot flow backwards!

Regardless of the concepts and how time flows, everyone uses it in one way or the other. For example, an American, who is immersed in a profit-oriented society, interprets time thus—time is truly money! Consequently, time is perceived as a precious, even scarce, commodity and to benefit from its passing, one has to move fast with it. So, for the action oriented American, therefore, there is no room for idleness. While they acknowledge that the past is over, they strive to seize the present, "parcel and package" it so they can make it work in the immediate future.

But can any one package time? Given the human perception of time, does it even have any meaning besides humans? Do plants, animals, and even the gods have any perception of time? Of course, we must also ask if time has any meaning for anything except us? There are many other questions we need to ask. For example, does color exist if it cannot be seen/perceived? Where and if we perceive color, what aspect of time is integral to that perception. At what point in *time* does the actual perception occur? Because in the Western reference time, has an arrow from past to present, to future, it can be measured. However, this measurement is unusual and reflects peculiar qualities of time because time here is measured by motion and actually becomes evident only through motion (Science daily: what is time? 2005)!

That means that *time*, whether measured as an instant or as an hour or as a day, or even as a year, is nothing without an event to mark it. Consequently, our sense of time is intimately influenced by the nature of the events themselves. Time also changes with perspective. This can be best explained by looking at the different ways in which time is categorized.

Yet, while the present gives us the most real feeling of time, almost all of what we perceive as now is already past. Therefore, the present is a fleeting moment! What is happening now (present) becomes so confined to an infinitesimally narrow point on the time line that we cannot even catch it because it is already being encroached upon by the past and the future!

Since the above is true when time is measured, can one really measure time? Or is time an illusion? The future appears to be a projection created by our past experiences stored in our memory. The fact that the present which gives us the most real feeling of time cannot be measured while the inaccessible past and future can be measured as durations suggests that the way we perceive time is an illusion. Hence, "People like us who believe in physics know that the distinction between the past, the present, and the future is only a stubbornly persistent illusion (What is time? 2018)."

## 1.3   Space

So, despite our differences and the cultural time, somehow, we accept the concept of time without giving it a second thought. But our collective thoughts do not imply that the universe is simply a great big clock. Each time we address the concept of time, we find that it is related to something else besides itself. For example, time relates to the swing of a pendulum; the vibrations of a quartz crystal; how the earth orbits; the motions of our limbs; the transportation of people, goods and services; the quantum leaping of atoms; the motions of magnetic and electric fields; the lives of different stars in the universe; the weddings of couples; the movement of electronic health data and even when we aggregate at the luncheon or dinner tables.

Therefore, and for all practical purposes, every idea of time that we can muster is deeply connected to a concrete physical event. Without such events, what would time consist of? Since time cannot exist in a void since it must relate to something—we also tag that something in relation to space!

What then is space? Space is a phenomenon that is differentially defined in science, mathematics and in communication (Space 2018). In the interest of *time* and *space* (used colloquially here), very brisk definitions are provided. Despite those different definitions in science, mathematics and digital communications, space is attributed with three dimensions of measurements. So, in astronomy and cosmology, for example, it is that huge 3-dimensional region that begins where the earth's atmosphere ends. Whereas in mathematics, space is an unbounded continuum (of unbroken set of points) for which any given point is defined by exactly three numerical coordinates (or distance dimensions) hence, sometimes called a 3-D space.

Finally, for digital communications, space refers to two things: first, it can refer to the interval during which no signal is transmitted, or during which the signal represents logic 0; or secondly it can be used in reference to the time interval that separates two characters, bytes, octets, or words in a digital signal. Collectively, then, space is the "boundless three-dimensional extent in which objects and events have relative position and direction." Physical space, although often conceived in three linear dimensions, when and where time is considered a dimension as it is by modern physicists, the result is sometimes called 4-space, 4-dimensional space, time-space, or space-time, or spacetime.

Spacetime in physics, therefore, is a mathematical model that fuses the three space dimensions with the time one dimension to produce one four-dimensional continuum (Wikipedia: Spacetime 2018). Figure 1.1 below is an artist's rendition of Gravity Probe B orbiting the Earth to measure space-time, a four-dimensional description of the universe that includes height, width, length, and time. Spacetime diagrams are helpful in visualizing relativistic effects as illustrated by the dictum that explains why different observers perceive *where* and *when* events occur.

However, does space have any meaning by itself? For example, if all the universe contained was a body of water, does this body of water have any position? The question becomes meaningless because, a position can only be defined with another

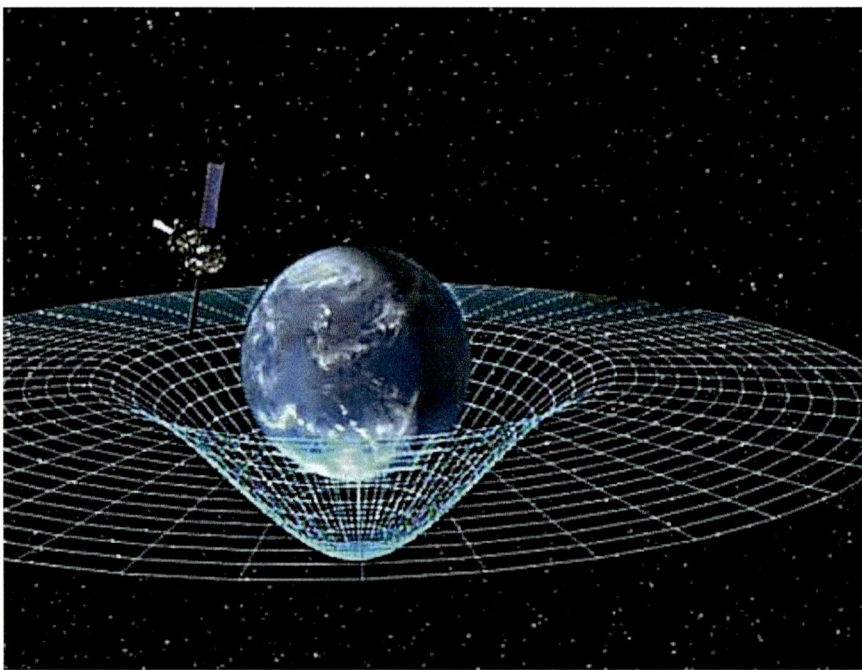

**Fig. 1.1** Artist concept of Gravity Probe B orbiting the Earth to measure space-time, a four-dimensional description of the universe including height, width, length, and time. (Reproduced from Wikipedia: Spacetime (2018)

position or thing. We cannot even ascribe a size to the body of water because there is nothing else to compare it to. Whether this body of water changes or not is dependent on whether something comes along and impacts it. But then, does this body of water even exist if it doesn't change? By the same token, we cannot tell the passage of time. Consequently, neither space nor time has any meaning whatsoever!

Remember Einstein's general theory of relativity? It posits that the continued development of space "may result in the collapse of the universe (Science daily: what is time? 2005)." That means that matter, as we know it, "would shrink into a tiny dot again (Science daily: what is time? 2005)." This would of course, end this concept of time that we have narrated (Science daily: what is time? 2005)! Of course, Enqvist of the University of Helsinki among others maintain that the "latest observations, however, do not support the idea of collapse, rather inter-galactic distances grow at a rapid pace (Science daily: what is time? 2005)."

While we have addressed an extreme scenario, the point being made here is that despite our differential (cultural included) perceptions of time and space, what really matters is how things, objects, people interact! It is simply because of all of the things that exist in the universe, that that the concept of time and space have meaning. It is this interaction that provides a framework in which time and space have meaning. This interaction is at the core of this book!

## 1.4 The Universe

"We live in a strange and wonderful universe. Its age, size, violence, and beauty require extraordinary imagination to appreciate. The place we humans hold within this vast cosmos can seem pretty insignificant (Hawking and Miodinow 2008)." The authors in the book, simply described what progress we humans have made in trying to find a unifying theory of all the forces of physics. That progress hardly touches the tip of the iceberg! The take away for us is that we still live in the age of discoveries, especially regarding the fundamental laws of nature.

## 1.5 Overview

The many and different chapters of this book try to present the different aspects of the healthcare ecosystem, that includes the electronic health/medical records. An attempt is made also to provide a direction that conserves time and space and thus reduces the escalating cost of a fragmented healthcare system. Every porting (movement) of an electronic health (medical record) involves time and space, the interaction of which results in a specified cost.

Because of the tremendous diversity and variations of today's EHRs (EMRs) it is difficult to pigeonhole them inside any standardized form. The costs of processing them have become incalculable, and the healthcare costs in the U.S. are spiraling out of control. Moreover, notions of these records vary across the globe. As incomplete and imprecise as they are, the U.S. definitions are a far cry from the European definitions. Yet they (in the U.S.) are governed by certain laws, whose limits are also notable. Within the array of emerging and mobile technologies, including the Internet of Things (IoT) (Eleven reasons why distributed ledger technology is perfect for healthcare 2017), both the US healthcare at large and the EHR (EMR) have found some strongholds. At the core of the different manifestations, is how the U.S. Healthcare and delivery systems lag behind those of other advanced countries. Yet technology and approaches which could reduce the lag and cost for the U.S. have evolved. The Blockchain technology is one such promising technology (Linn and Koo 2018; Blockchain in healthcare: a data-centric perspective 2018; Ribitzky et al. 2018; Shashank 2018). Commoditization is an approach that is also gaining grounds. These have been discoursed adequately in the book to warrant attention. They suggest future directions the US healthcare could take to lower costs (and possible) provide better care. At the core of such reduction of costs is the refinement of time and space interactions.

## 1.6   Time/Space Interaction

As already articulated, spacetime is the composite of the continuum of the three dimensions of space (length, width and height) and the one direction of time. Every creation of spacetime (using the definition in this book) creates a cost. The movement of EHR/EMR from point A to point B creates an interaction. Spacetime also place a huge role in interpersonal relations within any given society, although we are farther away from defining them precisely. So, then we descend to what humans understand and can explain (at least partially) by interactions. At a minimum and regardless of what the views are, when persons meet and interact, that period, that interval or that encounter results in an interaction. It becomes very precious, especially, if it involves a didactic relationship between a doctor and his/her patient. As fragmented as the EHR/EMR is, every patient, potentially carries part (if not all) of their health records.

And so, one of the challenges facing today's mobile society is how to capture this enriched encounter. In the medical and public health fields, we can capture the richness of the encounter (interaction) because we now become (since can we carry) the databases that constitute our health records. This enables us, therefore, as people or patients to engage in a rewarding and meaningful didactic relationship with the provider. It could be because that engagement is independent of the inhibitions of time and space, that we feel this new freedom to interact. It could be because, we are more empowered to engage intelligently in that encounter, that we strive to enrich the experience!

Critical in medicine and public health, therefore, are the interactions (deemed communications here) that ensue between a patient and the provider in any such un/defined time and space. Such an interaction is advocated in patient-centered deliveries of care. The fostering of such interaction is critical because it is one way to engender trust between the parties of that didactic relationship. It is also critical because it does encourage patients and persons to engage actively in not only their health, but in the decisions that affect their health. There are several derivatives of such engagements. From the patient's (person's) perspective are the relative importance of preventive and health maintenance, as well as compliance/adherence to treatment regimens, and consequently, the improvement of health outcomes.

A major derivative from the provider's perspective is that a good patient-provider communication allows the healthcare provider to provide better care. The provider can do so by better diagnoses of a patient's health condition because of better identification of the patient's health needs.

For these and many other reasons, we would like to mention some of the models that can be seen as anchors as well as benchmarks from which parameters can be drawn for measures. The models have been classified as:

Health Information Exchange (2016) which articulated these four effects on the provider regarding the timely sharing of vital patient information at the point of care: decrease duplicate testing; avoid medication errors; avoid readmissions and Improve decision making. These are fed into the electronic health data. Invariably then, the patient carries his or her own health database which can easily be ported into an electronic health record. The electronic exchange of health information subsumes, of course, the standardization of the data. In the context of this book, it also implies the aggregation (and or consolidation) of all patient data, a role that will be served by machine learning algorithm of AI. It also includes commoditization of patient records to enable them deliver at lower costs.

## 1.7 Conclusion

On balance then, neither time nor space would serve as a barrier for patient care. Because a patient is automatically carrying his or her health database, in a mobile form, and because the providers equally have access to the said database, interaction becomes the critical part of the engagement. Having now agreed upon manifestations of the escalating costs—from the space in which electronic health records are stored, to the essential manifestations of escalating costs due to the times of execution, spatial and temporal components should also, be critical as electronic health records are ported. (Of course, the inability to provide the exact costs in any space-time dimensions obliterates the actual cost of both time and space.) But fragmented sources of EHR augur well for differential and sometimes intractable costs. Blockchain technology would allow electronic health records (EHR) to be ported without the restrictions of interoperability, data security, and person's privacy breaches. Incentivized EHR usage yields extraordinarily Big data whose analytics creates an extra layer of burden. Continued fragmentation of health records increases the costs of healthcare. Fragmented health records also include the individual health records persons carry throughout their lives. Capturing these during didactic interactions between doctors and patients and weaving these into the electronic health records would help consolidate a given patient's health record. The commoditization of EHR (and other health IT tools) would ameliorate the rising costs in the delivery of the US Healthcare (Lichtenwald 2017). These all involve the interaction of space and time whose incessant occurrences escalate the cost of healthcare.

However, moving forward, the commoditization of Healthcare, Blockchain technology and the extraction of salient health record data with AI machine learning would make the interaction of time and space to occur less frequently. These would reduce costs as they would occlude the following hindrances: of data exchange due to interoperability limitations; data exchange errors (both semantic and syntactic); security of data; and the privacy of the patient. In other words, spacetime—the interplay of space and time become both a necessary and a sufficient condition! This interplay would affect the cost of care, by streamlining it, reducing it, or completely eliminating it!

Additionally, and critical to all the involved entities (patient and providers—including doctors, pharmacists, nurses and other health care providers) is the fact that the digital exchanges (porting) of the patient health information potentiates improvements in different areas, including: speed, quality, safety and cost of patient care. In aggregate, the outcomes should provide rich data for data analytics—either on an individual, group or societal level(s)!

So, we invite you to this new path, where time and space interference are both crucial and obliterated. We invite you to the future of Portable Health Records in a Mobile society! Come join the group!

# References

Blockchain in healthcare: a data-centric perspective. 2018. https://medium.com/crypto-oracle/blockchain-in-healthcare-a-data-centric-perspective-109e898d73f3. Accessed 6 Sept 2018.

Christoforou P. 5 bizarre paradoxes of time travel explained. 2014. http://www.astronomytrek.com/5-bizarre-paradoxes-of-time-travel-explained/. Accessed 2 Sept 2018.

Clarke AC. IEEE Technology Time Machine: beyond tomorrow. 2018. http://sites.ieee.org/ttm/previous/ttm-2016/. Accessed 2 Sept 2018.

Eleven reasons why distributed ledger technology is perfect for healthcare. 2017. https://medium.com/crypto-oracle/eleven-reasons-why-distributed-ledger-technology-is-perfect-for-healthcare-bc6adb02406f. Accessed 2 Sept 2018.

Hawking S, Miodinow L. A briefer history of time. New York: Random House; 2008. p. 1.

Howell E. Is time travel possible? Scientist explore the past and the future. 2018. https://www.space.com/40716-time-travel-science-fiction-reality.html. Accessed 2 Sept 2018.

Lichtenwald I. HIT think why commodization has yet to affect healthcare IT. 2017. https://www.healthdatamanagement.com/opinion/why-commoditization-has-yet-to-affect-healthcare-it. Accessed 6 Sept 2018.

Linn LA, Koo MB. Blockchain for health data and its potential use in health IT and health care related research. 2018. https://www.healthit.gov/sites/default/files/11-74-ablockchain-forhealthcare.pdf. Accessed 2 Sept 2018.

May RM. Science and society. Science. 2001;292(5519):1021.

Ribitzky R, St. Clair J, Houlding DI, McFarlane CT, Ahier B, Gould M, et al. Pragmatic, interdisciplinary perspectives on Blockchain and distributed ledger technology: paving the future of healthcare. 2018. https://blockchainhealthcaretoday.com/index.php/journal/article/view/24/37. Accessed 6 Sept 2018.

Science daily: what is time? 2005. https://www.sciencedaily.com/releases/2005/04/050415115227. htm. Accessed 2 Sept 2018.

Shashank A. 5 benefits of using Blockchain technology in healthcare. 2018. https://hitconsultant. net/2018/01/29/blockchain-technology-in-healthcare-benefits/. Accessed 6 Sept 2018.

Space. 2018. https://whatis.techtarget.com/definition/space. Accessed 2 Sept 2018.

SpaceTime. Relativity, and quantum physics. 2017. http://www.ws5.com/spacetime/. Accessed 2 Sept 2018.

Stange KC. The problem of fragmentation and the need for integrative solutions. Ann Fam Med. 2009;7(2):100–3.

Tick-Tock. What is time? 2017. http://waleg.com/various/tick-tock-time/. Accessed 2 Sept 2018.

The Office of the National Coordinator for Health Information Techonology. https://www.healthit. gov/sites/default/files/ltpac_value_prop_factsheet_6-21-16.pdf

What is time? What is the cause of time? https://www.timephysics.com/. Accessed 2 Sept 2018.

Wikipedia: Spacetime. 2018. https://en.wikipedia.org/wiki/Spacetime. Accessed 2 Sept 2018.

Wikipedia: time in physics. 2018. https://en.wikipedia.org/wiki/Time_in_physics. Accessed 2 Sept 2018.

# Part II
# Health Records

# Chapter 2
# Health Records

**Solomon N. Koppoe**

**Abstract** A mobile society and the portability of our health records resulted in a change in the way our health records are kept and managed. Electronic Health Records (EHRs), Electronic Medical Records (EMRs) and Personal health records (PHRs) are different forms of electronic records kept for medical purposes. EHRs are built to go beyond standard clinical data collected in a provider's office and are inclusive of a broader view of a patient's care. EMRs were the first electronic sources used to electronically store patient information (mostly electronic versions of the paper charts). PHRs contain the same types of information as EHRs but may contain patient contributed information and are designed to be set up, accessed, and managed by patients. More complex PHR systems are now being integrated into health provider information systems, combining personal record keeping, access to current electronic health records, and a range of information and communication functions. EHR systems can include many potential capabilities, but three functionalities hold great promise in improving the quality of care and reducing costs at the health care system level: clinical decision support (CDS) tools, computerized physician order entry (CPOE) systems, and health information exchange (HIE). A CDS system is one that assists the provider in making patient care decisions. Using a CPOE system linked to a CDS, can result in improved efficiency and effectiveness of care. HIEs can help exchange patient information among providers to better facilitate patient care. EHRs have clinical, organizational, and social benefits. Clinical outcomes focus on the concept of quality in relation to direct patient-care, services, and treatments. Adoption of national standards will contribute to interoperability, trans- portability, and security features of EHRs and PHRs. As both EHRs and PHRs become standardized, patients will be able to move from one place to another and have their medical records accessible and transferable wherever they go.

**Keywords** Health · Records · Medical · Electronic · System · Patient · Physician Clinical · Access

S. N. Koppoe (✉)
Freemont Consulting, L.L.C., Gainesville, VA, USA
e-mail: solomon.koppoe@octoconsulting.com

© Springer Nature Switzerland AG 2019
E. R. Onyejekwe et al. (eds.), *Portable Health Records in a Mobile Society*,
Health Informatics, https://doi.org/10.1007/978-3-030-19937-1_2

## 2.1 Health Records

Health records can be maintained in traditional paper or modern electronic form. Paper-based records are generated by different medical entities (Amatayakul 2007). They have communication gaps which lead to repeat testing and treatments that are avoidable. Paper records involve the physical transfer of records by patients, mail or fax, and are time intensive (Amatayakul 2007) and subject to physical damage or loss. New providers must retrieve patients' physical records from multiple offices. The lack of centralized storage and management makes it difficult to establish a complete patient history (Amatayakul 2007). Additionally, doctors' access to medical records is limited by location and office hours; this can impact patient health in unusual circumstances, such an emergency or when vital medication is misplaced (Coffey et al. 2015).

Electronic Health Records (EHRs), Electronic Medical Records (EMRs) and Personal health records (PHRs) are different forms of electronic records. Providers and vendors sometimes use these terms interchangeably, but they are not the same (Coffey et al. 2015). EMRs and EHRs have become essential medium of communicating vital patient information to other members of the health care team as well as to patients (McMullen et al. 2014). EHRs are built to go beyond standard clinical data collected in a provider's office and provide a broader view of a patient's care. EHRs contain information from all the clinicians and allied health providers involved in a patient's care. All authorized clinicians involved in a patient's care can access the electronically stored information to help guide patient care. Most EHRs can also share information with unaffiliated health care providers, such as laboratories and specialists. EHRs can follow patients—to the specialist, the hospital, the nursing home, or even across the country (Amatayakul 2009).

EHRs have the potential to accommodate all of the functions of the EMR as well as important clinical information from an extended set of care providers. Typically, EMRs are associated with independent practices and EHRs are associated with Integrated Delivery Systems (IDSs). EHRs are specifically designed for sharing information among various types of providers who may be located in a number of settings (primary care, in-patient, emergency department, abroad) and between providers and patients. In a comprehensive EHR system, patients can log onto their own records, read and track test results, communicate with providers, and contribute information that ultimately improve their health (Garrett and Seidman 2011).

Historically, EMRs were the first electronic systems used to store patient information (Garrett and Seidman 2011) electronically. EMRs are digital versions of the traditional paper charts used in clinician offices, clinics, and hospitals. They are records maintained to meet the needs of a specific facility or organization. EMRs contain notes and information collected by and for the clinicians in that office, clinic, or hospital. They are mostly used by providers for diagnosis and treatment. EMRs are more valuable than paper records because they enable providers to analyze data over time, identify patients for preventive visits, screenings, monitoring, and improve health care quality (Amatayakul 2009). EMRs grew in popularity because of the added benefits not present in paper charts, including the ability to easily collate and track sets of information, monitor changes in patient outcomes after implementation of a new practice or procedure, and determine which patients are due for physical exams, procedures, immunizations, and the like. Unfortunately, EMRs are often practice-specific, making it difficult to transfer information to outside groups of providers, to other health care systems, and to patients (Coffey et al. 2015; McMullen et al. 2014). The use of an EMR is generally considered to be superior to paper documentation because they are more accessible, more legible, and automatically author, time, and date stamp the documentation (Eastes et al. 2010).

However, electronic documentation is often perceived to be more cumbersome than paper because it requires more involved hand-eye coordination skills (i.e., inputting data via a keyboard and mouse rather than using a paper and pen) and more cognitive energy to navigate and interact with data fields on a screen compared with the more familiar layout of paper documentation. Because of the complex and rapid nature of trauma resuscitations, missing data elements is a common problem in trauma documentation (e.g., flow sheets (Coffey et al. 2015)). These data elements may be critical to patient care and safety (e.g., intravenous fluid volumes, vital signs, and primary and secondary assessment) or critical to trauma registry and performance improvement (e.g., physician arrival time, injury mechanism, and disposition from the emergency department). Most trauma centers continue to use handwritten flow sheets to document trauma resuscitations because of concerns about thoroughness and timeliness of data collection (Coffey et al. 2015). One trauma center was warned against adopting electronic trauma flow sheet use in the trauma room during a recent site survey, citing concerns that the American College of Surgeon reviewers would not be able to navigate through the EMR, which could put the center at risk during their reverification process (Eastes et al. 2010).

PHRs contain the same types of information as EHRs—e.g. diagnoses, medications, immunizations, family medical histories, and provider contact information—but are designed to be set up, accessed, and managed by patients. Patients can use PHRs to maintain and manage their health information in a private, secure, and confidential environment. PHRs can include information from a variety of sources including clinicians, home monitoring devices, and patients themselves (Amatayakul 2009).

Personal Health Records (PHRs)  have emerged as one of the solutions to the increasing demand of patients for flexible access to health information and services (Kaelber et al. 2008). PHRs are described as patient-centered records which allow pertinent medical information to be collected, organized and maintained at least in part by the individual patient (Kaelber et al. 2008). PHRs are also described as "Electronic personal health record (PHR): a private [and] secure application through which an individual may access, manage, and share his or her health information. They include information that is entered by the consumer and/or data from other sources such as pharmacies, labs, and health care providers. The PHR may or may not include information from the EHR (Jones et al. 2010). PHR sponsors include vendors who may or may not charge a fee, health care organizations such as hospitals, health insurance companies, or employers. In addition to copies of selected EHR data, PHRs comprise patient generated health and lifestyle records that are stored and managed using a personal computer or web application linked to provider held records (Iakovidis 1998). More complex PHR systems are now being integrated into health provider information systems, combining personal record keeping, access to current electronic health records, and a range of information and communication functions. Advanced features in the early developmental stages include appointment scheduling, prescription renewals, medical history question-naires, remote medical visits (e-visits), and access to patient specific medical litera-ture, where patients can review databases and research relevant health conditions (Kaelber et al. 2008; Maloney and Wright 2010). Patients are increasingly searching for more accessible and portable options for maintaining and accessing their per-sonal health records which has resulted in the development of PHRs stored on elec-tronic personal devices (Maloney and Wright 2010).

Mobile PHRs included USB drives, CDs, and other electronic storage devices that were incorporated into bracelets or wallet cards (Kharrazi et al. 2012). The primary function of many of these portable devices was to provide critical medical history information to providers in times of emergency, and marketing of these devices was driven by scare tactics and scenarios where lack of medical information could result in serious injury or death (Maloney and Wright 2010). These early, portable, personal electronic devices had significant limitations, including insuffi-cient security safeguards and lack of interoperability, rendering them useless if the medical data could not be accessed (Office of the National Coordinator for Health Information Technology 2016). Moreover, these devices are not stand-alone and require an external computer to read the data from the portable PHR, or contain supplemental proprietary software needed to access the medical data.

As technology becomes increasingly portable and interactive, cellular phones and tablet computers have emerged as a new potential platform for PHRs (Kharrazi et al. 2012). The use of cell phones has dramatically increased, and individuals are becoming more technologically savvy with the availability of "smart phones." Almost all aspects of computer and personal use are being integrated into "smart phone" applications, and new health-related software is also being developed (Coldewey 2010). PHRs accessible to "smart phones" are a natural progression from the current mobile PHR storage. Mobile Personal Health Records (PHRs), sometimes called mobile health or mHealth, are mobile phone or PDA applications that help you track your health. Mobile PHRs can be used with any cellular phone with Internet access. Hundreds of applications have already been developed to address specific health needs. Mobile PHR apps can help monitor health conditions, track medication schedules, locate a hospital or doctor, stick to a healthier lifestyle, and increase access to more health information (Kharrazi et al. 2012).

## 2.2  Purpose

EHRs include patient demographics, progress notes, problems, care plans, medica-tions, vital signs, past medical history, immunizations, laboratory data, and radiol-ogy reports (Menachemi and Collum 2011). Some of the basic benefits associated with EHRs include being able to easily access computerized records and eliminate poor penmanship, which has historically plagued the medical chart (Rodríguez-Vera et al. 2002). EHR systems include many capabilities, but three particular functionalities hold great promise in improving the quality of care and reducing costs at the health care system level: clinical decision support (CDS) tools, computerized physician order entry (CPOE) systems, and health information exchange (HIE).

A CDS system is one that assists the provider in making decisions with regard to patient care. Some functionalities of a CDS system include providing the latest information about a drug, cross-referencing a patient allergy to a medication, alerts for drug interactions, cost considerations, duplication and other potential patient issues that are flagged by the computer. With the continuous growth of medical knowledge, each of these functionalities provides a means for care to be delivered in a much safer and more efficient manner (Menachemi and Collum 2011).

CPOE systems allow providers to enter orders (e.g., for drugs, laboratory tests, radiology, physical therapy) into a computer rather than doing so on paper. Computerization of this process eliminates potentially dangerous medical errors caused by poor physician handwriting. It also makes the ordering process more efficient because nursing and pharmacy staffs do not need to seek clarification or to solicit missing information from illegible or incomplete orders. Using a CPOE system, especially when it is linked to a CDS, can result in improved efficiency and effectiveness of care (Bates et al. 1998).

EHRs facilitate the sharing of patient information through HIE, which is the process of sharing patient-level electronic health information between organizations and can create efficiencies in the delivery of health care (Walker et al. 2005). By allowing for the secure sharing of patient information, HIE can reduce costly redundant tests that are ordered because one provider does not have access to the clinical information stored at another provider's location.

Patients typically have data stored in a variety of locations where they receive care (Menachemi and Collum 2011). This can include their primary care physician's (PCP) office, physician specialists, one or more pharmacies, hospitals, emergency departments and other patient care locations. Over a lifetime, large amounts data can accumulate at a variety of places, creating isolated silos of information. Historically, providers rely on faxing or mailing copies pertinent patient records that makes it difficult and time consuming to access when and where needed. HIE facilitates the electronic exchange of patient records between EHRs, which can result in more cost-effective and higher-quality care (Menachemi and Collum 2011).

The advantages of an EHR can be grouped into clinical; organizational; and social benefits (Menachemi and Collum 2011). Clinical outcomes focus on the concept of quality in relation to direct patient-care services and treatments. Studies have confirmed that EHRs can increase the quality of patient-care services and treatments by serving as a platform for timely access to complete and accurate patient information (McMullen et al. 2014). Such information can be used to support health care providers in care planning, care delivery, and the monitoring of patient responses to the services and treatments provided (McMullen et al. 2014). Communication of important health information, such as current medication use, allergies, health history, and other data are a valuable tool in reducing adverse clinical events because the information can be used to alert providers to critical patient information (Institute of Medicine 2003).

Organizational entities within a health care facility, such as health information management, case management, and population health management, are EHR beneficiaries as well. Utilization of EHRs typically increases medical billing and coding accuracy, improves rates of reimbursement from third-party payers, increases job productivity and satisfaction among direct and indirect users of the EHR, and results in a decline in medical errors (Menachemi and Collum 2011). EHRs not only improve the quality of care, but such systems can also reduce health care costs by improving outcomes, resulting in better management of chronic illnesses, and reduce the duplication of services (Menachemi and Collum 2011). EHRs facilitate research by collecting data that can then be collated into larger data sets, leading to more powerful quantitative research studies, the findings of which are more generalizable to other patient situations. Additionally, studies have demonstrated that adoption of EHRs improves provider satisfaction, likely due to such factors as ease of access to information, faster charting times (once the system has been mastered), and retrieval of information from multiple sources (Menachemi and Collum 2011).

EHRs not only affect providers and health care agencies, but they support the patients' ability to follow their own care plans and insure that the information is available to those designated by the patient, whether it be a "significant-other" or a health care provider. EHRs also facilitate a patient's ability to review information contained in the record, absorb medical information at their own pace, question what is not understandable, provide additional information that has not been solicited, and report additional information concerning activities that lead to a healthier lifestyle, such as joining a health club, receipt of acupuncture, or new membership in a weight-management plan (Institute of Medicine 2003). A recent study was conducted by Reed and colleagues to determine whether utilization of an EHR system could positively impact health outcomes among over 169,000 patients with diabetes. Study participants who had access to their health care information demonstrated significant improvements in their hemoglobin $A_{1c}$ values, lipid levels, and frequency of monitoring, particularly among those whose diabetes was not previously well controlled (Reed et al. 2012).

## 2.3 Limits

Despite the growing literature on benefits of various EHR functionalities, some authors have identified potential disadvantages associated with this technology. These include financial issues, changes in workflow, temporary loss of productivity associated with EHR adoption, privacy and security concerns, and several unintended consequences. Financial issues, including adoption and implementation costs, ongoing maintenance costs, loss of revenue associated with temporary loss of productivity, and declines in revenue, present a disincentive for hospitals and physicians to adopt and implement an EHR. EHR adoption and implementation costs include purchasing and installing hardware and software, converting paper charts to electronic ones, conversion of data from legacy systems, integration with other operational systems, and end-user training. Many studies have documented these costs in both the inpatient and outpatient settings (McMullen et al. 2014; Agrawal 2002). Another disadvantage of an EHR is disruption of work-flows for medical staff and providers, which results in temporary losses in productivity (Hripcsak and Albers 2012). This loss of productivity stems from end-users learning the new system and frequently leads to losses in revenue (Menachemi and Collum 2011). Another potential drawback of EHRs is the risk of patient privacy violations, which is an increasing concern for patients due to the increasing amount of health information held and exchanged electronically (Amatayakul 2011).

Viable EHR systems must constantly work to prevent unauthorized patient information access that may originate from internal and external pathways (Hripcsak and Albers 2012). Internal threats to private patient information result from poor password management, disgruntled and disloyal employees, or ineffective physical security measures. External threats include unauthorized access to protected health information by malicious actors and theft or loss of electronic devices containing health information (Hripcsak and Albers 2012; Amatayakul 2011). To mitigate some of these concerns, policymakers have taken additional measures to ensure safety and privacy of patient data. Recent legislation has imposed regulations specifically relating to the electronic exchange of health information that strengthen existing Health Insurance Portability and Accountability Act privacy and security policies (Parver 2009). Although few electronic information systems are completely secure, the rigorous requirements set forth by new legislation make it much more difficult for electronic data to be accessed inappropriately (Hripcsak and Albers 2012). EHR systems are required to have an audit function that allows system operators to identify individuals who accessed various aspects of a given medical record (Menachemi and Collum 2011). Many hospitals and physicians are implementing strict, no tolerance penalties for employees who access patient information inappropriately. A hospital in Arizona terminated several employees after they inappropriately accessed the records of victims who were hospitalized after the January 2011 shooting involving a US Congresswoman (Innes 2011).

EHRs may cause several unintended consequences, such as increased medical errors, negative emotions, changes in power structure, and overdependence on technology. Researchers have found an association between the use of CPOE and increased medical errors due to poorly designed system interfaces or lack of end-user training (Campbell et al. 2006).

## 2.4   Content

Each member or patient is assigned a unique medical record, which contains at least the following information:

The identification of a Primary Care Physician (PCP) who coordinates care, where the member's plan requires a PCP assignment, the record verifies the PCP that coordinates and manages the member's care (Emblem Heal n.d.).

Personal demographic records which include name, id number, date of birth, address and phone number, employer's name, address and phone number, marital status, benefit plan participation and copayment (if applicable), name of the primary care physician (PCP) , list of allergies and/or adverse reactions, or "No Known Allergies" (NKA) (Emblem Heal n.d.).

Medical history that contains biographical information, comprehensive baseline history, physical observations, diagnostic test results, consult reports, progress notes, medication records, problem list, allergy documentation, telephone/communication log, immunization records, preventive health screening records, inpatient/ER discharge summary reports and operative reports (Emblem Heal n.d.). The PCP must also clearly document any follow up on the member's ER visits and/or hospitalizations, whether from an office visit, written correspondence, or telephone conversation (Emblem Heal n.d.).

The comprehensive baseline history and physical assessment must include a review of subjective and objective complaints/problems, family history, social history (i.e., occupation, education, living situation, risk behaviors), significant accidents, surgeries, illnesses and mental health issues, complete and comprehensive review of systems (including patient's presenting complaint, if applicable), prenatal care and birth information (baseline, 18 years and younger only). In cases where the member has both a PCP and an OB/GYN, care must be coordinated to ensure there is a centralized medical record for the provision of prenatal care and all other services (Emblem Heal n.d.). Periodic history and physical review should be repeated in accordance with age appropriate preventive care guidelines (Emblem Heal n.d.). Within the record jacket, reports of similar type (i.e., progress notes, laboratory reports) should be filed together in chronological or reverse chronological order permitting easy retrieval of information and initialed by the physician to indicate they have been read (Emblem Heal n.d.). Each progress note should be legibly written or typed, signed, and dated by the author. It should contain at least the following items: reason for visit as stated by the member, the duration of the problem, findings on physical examination, any laboratory and X-ray results, diagnosis or assessment of the member's condition, and any therapeutic or preventive services prescribed. In terms of medications, there should be clear documentation of dosage, duration and side effect information of any prescription given. Medication allergies and adverse reactions must be noted prominently (updated during a physical, when a prescription is written, or annually, whichever comes soonest), along with a follow-up plan (including self-care training) or clarification that no follow up is required (Emblem Heal n.d.).

## 2.5  Standards

PHRs may be certified. The Certification Commission for Health Information Technology (CCHIT) has both a Personal Health Record Work Group and a PHR Advisory Task Force, the latter of which has recommended certification of these PHR attributes: privacy, security, interoperability, and functionality (Certification Commission for Healthcare Information Technology 2008). Adoption of national standards will be important for interoperability, trans-portability, and security, features that will likely be mandated by legislation (Certification Commission for Healthcare Information Technology 2008). As both EHRs and PHRs become standardized, patients will be able to move from one provider to another and have their medical records accessible and transferable wherever they go (Certification Commission for Healthcare Information Technology 2008). MyChart, developed by Epic, is one of the most widely used PHRs by health systems such as Kaiser Permanente. Other efforts by Apple, Amazon, Chase, and Berkshire Hathaway show promise, as well.

# References

Agrawal A. Return on investment analysis for a computer-based patient record in the outpatient clinic setting. J Assoc Acad Minor Phys. 2002;13(3):61–5.

Amatayakul M. Electronic health records: a practical guide for professionals and organizations. Chicago: American Health Information Management Association; 2007.

Amatayakul M. EHR versus EMR: what's in a name? Healthcare Financial Management. 2009;63(3):24–5.

Amatayakul M. Why workflow redesign alone is not enough for EHR success: addressing workflow changes has been heralded as vital to electronic health record (EHR) success. Undoubtedly, workflows will change with an EHR and should be anticipated, planned for, and carefully implemented. Healthc Financ Manage. 2011;65(3):130–2.

Bates DW, Leape LL, Cullen DJ, Laird N, Petersen LA, Teich JM, et al. Effect of computerized physician order entry and a team intervention on prevention of serious medication errors. JAMA. 1998;280(15):1311–6.

Campbell EM, Sittig DF, Ash JS, Guappone KP, Dykstra RH. Types of unintended consequences related to computerized provider order entry. J Am Med Inform Assoc. 2006;13(5):547–56.

Certification Commission for Healthcare Information Technology. Recommendations of the PHR Advisory Task Force: certification of PHRs [Internet]. The Commission; 2008 [cited 27 Aug 2009]. http://www.hitanalyst.files.wordpress.com/2008/07/cchitphratf.pdf.

Coffey C, Wurster LA, Groner J, Hoffman J, Hendren V, Nuss K, et al. A comparison of paper documentation to electronic documentation for trauma resuscitations at a level I pediatric trauma center. J Emerg Nurs. 2015;41(1):52–6.

Coldewey D. Mobile crunch. 2010. http://www.mobilecrunch.com/2010/09/15/pew-survey-finds-predictable-trends-among-mobile-phone-users.

Eastes LE, Johnson J, Harrahill M. Using the electronic medical record for trauma resuscitations: is it possible? J Emerg Nurs. 2010;36(4):381–4.

Emblem Heal. Medical record content and format. https://www.emblemhealth.com/providers/provider-manual/medical-record-guidelines/content-and-format.

Garrett D, Seidman J. EMR vs EHR—what is the difference? Office of the National Coordinator for Health Information Technology (ONC). Updated January 4, 2011. http://www.healthit.gov/buzz-blog/electronic-health-and-medical-records/emr-vs-ehr-difference. Accessed 12 May 2014.

Hripcsak G, Albers DJ. Next-generation phenotyping of electronic health records. J Am Med Inform Assoc. 2012;20(1):117–21.

Iakovidis I. Towards personal health record: current situation, obstacles and trends in implementation of electronic healthcare record in Europe. Int J Med Inform. 1998;52(1-3):105–15.

Innes S. 3 UMC workers fired for invading records. Arizona Daily Star. 13 Jan 2011.

Institute of Medicine. Key capabilities of an electronic health record system: letter report. Washington, DC: National Academy Press; 2003. http://www.iom.edu/Reports/2003/Key-Capabilities-of-an-Electronic-Health-Record-System.aspx. Accessed 1 Aug 2014.

Jones DA, Shipman JP, Plaut DA, Selden CR. Characteristics of personal health records: findings of the Medical Library Association/National Library of Medicine Joint Electronic Personal Health Record Task Force. J Med Libr Assoc. 2010;98(3):243–9.

Kaelber DC, Jha AK, Johnston D, Middleton B, Bates DW. A research agenda for personal health records (PHRs). J Am Med Inform Assoc. 2008;15(6):729–36.

Kharrazi H, Chisholm R, VanNasdale D, Thompson B. Mobile personal health records: an evaluation of features and functionality. Int J Med Inform. 2012;81(9):579–93.

Maloney FL, Wright A. USB-based Personal Health Records: an analysis of features and functionality. Int J Med Inform. 2010;79(2):97–111.

McMullen PC, Howie WO, Philipsen N, Bryant VC, Setlow PD, Calhoun M, Green ZD. Electronic medical records and electronic health records: overview for nurse practitioners. J Nurse Pract. 2014;10(9):660–5. https://doi.org/10.1016/j.nurpra.2014.07.013.

Menachemi N, Collum TH. Benefits and drawbacks of electronic health record systems. Risk Manag Healthc Policy. 2011;4:47–55. https://doi.org/10.2147/RMHP.S12985.

Office of the National Coordinator for Health Information Technology. Percent of hospitals, by type, that possess certified health IT. Health IT Quick-Stat #52. May 2016. https://dashboard.healthit.gov/quickstats/pages/certified-electronic-health-record-technology-in-hospitals.php. Last visited 27 Feb 2017.

Parver C. How the American Recovery and Reinvestment Act of 2009 changed HIPAA's privacy requirements. CCH Health Care Compliance Letter. 28 July 2009. pp. 4–7.

Reed M, Huang J, Graetz I, Brand R, Hsu J, Fireman B, Jaffe M. Outpatient electronic health records and the clinical care and outcomes of patients with diabetes mellitus. Ann Intern Med. 2012;157(7):482–9.

Rodríguez-Vera FJ, Marin Y, Sanchez A, Borrachero C, Pujol E. Illegible handwriting in medical records. J R Soc Med. 2002;95(11):545–6.

Walker J, Pan E, Johnston D, Adler-Milstein J. The value of health care information exchange and interoperability. Health Aff. 2005;24:W5–10.

# Chapter 3
# Standards

Jefferson L. Howe

**Abstract**  As health care delivery changed and the volume and complexity of treatments advanced, so did the requirements and guidelines for medical record documentation. Comprehensive health documentation management in today's modern culture involves adherence to a multitude of Federal and State regulatory requirements, compliance with various accreditation bodies, and professional practice standards, and organizationally developed policies. Together these requirements work to result in comprehensive health record documentation that includes information pertinent to the care and treatment of the patient in order to promote continuity of care, justify the care that was rendered, and provide evidence for medical necessity, patient education, billing compliance, and defense against litigation. Today, this effort presents an ever-increasing need for stewardship and integrity for the information that is gathered so as not to proliferate data in the absence of sound treatment information.

**Keywords**  EHR standards · Meaningful use · HITECH Act

Health record standards have continually evolved to meet the needs of the changing landscape of healthcare since 1928 when the American College of Surgeons (ACS) established the Association of Record Librarians of North America in order to "elevate the standards of clinical records in hospitals, dispensaries, and other distinctly medical institutions (Spath 2009)." This was perhaps the first authority to establish a general model for medical record practices. The ACS in its *Manual of Hospital Standards* provided guidelines pertaining to documentation in records along with the maintenance and use of medical record information. This created the foundation for what has advanced to modern health information management (Spath 2009). Today, the established requirements for health records in technical systems

J. L. Howe (✉)
Department of Health Information Integrity, University of Miami Health System,
Miami, FL, USA
e-mail: Jefferson.Howe@gc.stephens.edu

© Springer Nature Switzerland AG 2019
E. R. Onyejekwe et al. (eds.), *Portable Health Records in a Mobile Society*,
Health Informatics, https://doi.org/10.1007/978-3-030-19937-1_3

originates from the basic concept that health records serve as the only evidence of treatment and services rendered. As such, each patient visit should result in documentation and information capture that collects core elements meant to positively identify the patient, establish presenting problems as well as subjective and objective findings that culminate in the establishment of a diagnosis or primary condition. In addition, documentation of treatment along with recommendations and plans should also be part of the healthcare encounter.

Electronic Health Records and rapid expansion of information technology has created challenges given that historical guidelines may compete with electronic integration and the source of information may be difficult to ascertain.

## 3.1 Federal Standards

The Meaningful Use regulation promotes a federal standard for electronic health records that is meant to enhance the adoption of use while at the same time providing efforts to improve quality and efficiency of healthcare. Meaningful Use for Electronic Health Records (EHR), implemented in 2011 as a part of the Health Information Technology for Economic and Clinical Health Act (HITECH) advanced the adoption of electronic health record utilization by establishing elements of "meaningful" use that would demonstrate the effectiveness of applying information technology to traditional record keeping and treatment practices. Under HITECH, Meaningful Use standards offered specific electronic use objectives that if demonstrated and used would result in payment incentives to eligible physicians and hospitals treating Medicare and Medicaid patients. Primary objectives include basic data entry of patient medical record information including vital signs, demographics, medications, allergies, up-to-date problem lists of current and active diagnoses and smoking status (Blumenthal and Tavenner 2010). Once the basic elements are captured at each treatment visit, these core elements can combine with software applications that help to improve patient safety by comparison using clinical decision support tools. There are a number of potential performance measurement functions along with a number of potential patient engagement functions that represent more advanced functionality of the EHR.

The efforts in complying with the Meaningful Use criteria have resulted in higher levels of adoption of electronic health record technology across the industry. The *Journal of the American Medical Informatics Association* reported "a total of 80.5% of US hospitals had adopted a basic EHR by 2015, an increase of 5.3% points from 2014 (75.2%) (Adler-Milstein et al. 2017)." The article stated that 100% adoption would be possible in the next 4 years. The Office of the National Coordinator for Health Information Technology provides a dashboard with EHR adoption rates, which is updated as per reported data. Moving forward, it will be important for the industry to establish a baseline for functionality in order to assess progress.

The Health Insurance Portability and Accountability Act (HIPAA) of 1996 has had significant impact on the advancement of electronic health records and standards by resulting in regulations that protect the "privacy of health information in which the patient is identifiable (Annas 2003)." The results have been far-reaching in scope, causing HIPAA to impact virtually all areas of the United States health care system. Becoming effective in 2003, the HIPAA provisions outline standards to be undertaken in the transmission and disclosure of protected health information. The rule is broad in scope and applies to "health plans, health care clearinghouses and health care providers (hospitals, clinics, and health departments) who conduct financial transactions electronically ("covered entities") (Gostin 2001). In most cases, an expressed patient authorization is required to release protected health information unless the release is for treatment, payment, or hospital operations. The rule applies to identifiable information in any form whether communicated electronically, on paper, or orally.

In addition, HIPAA allows the patient the right to notice of privacy practices along with consumer access to medical records. The intent is to provide transparency in the use and exchange of information. Providers and health plans must give patients a written explanation of "allowable uses and disclosures of protected health information and patients' rights (U.S. Government Publishing Office: 45 CFR 164.520 2002)."

These standards have been a long topic of national debate and anecdotal evidence suggests that HIPAA provisions may have resulted in slowing the adoption of electronic health record utilization while organizations and technologies grappled to understand the requirements. Today, despite advancements in integration of electronic information exchange, there is still a preponderance of medical information that is shared on paper by printing and faxing distribution.

Most initiatives and advancements promulgated through federal regulations are meant to protect patient privacy, increase patient safety, advance healthcare delivery, protect billing and improve compliance, or provide justification for medical necessity. Such initiatives often result in the proliferation of medical record documentation in order to substantiate the effort. Medical information can be documented and captured primarily by physicians and nursing professionals. In addition, other allied health professionals, physical therapists, occupational therapists, pharmacists, dietitians, and speech therapists may document in the health record. Efforts to adhere to regulations may result in increasing documentation responsibilities for the healthcare professionals.

## 3.2 State Regulatory Requirements

Since hospitals and healthcare agencies are licensed by the state in which they operate, often Federal Regulation requirements are delegated to the state's licensing agency to monitor compliance and effectiveness. As a result, individual states may develop independent guidelines or regulatory requirements that augment documentation specifications in the oversight of patient care. This may be more common for state-operated psychiatric facilities, for example, that have specific provisions for confidentiality. This may also be common in state-defined requirements for emergency medical treatment under the Emergency Medical Treatment and Labor Act (EMTALA). EMTALA requires hospitals with emergency departments to provide a "medical screening examination to any individual who comes to the emergency department and requests such an examination, and prohibits hospitals with emergency departments from refusing to examine or treat individuals with an emergency medical condition (Centers for Medicare and Medicaid Services 2012)." As a result, organizations in different states may have specific documentation requirements, or data elements, needed to comply.

## 3.3   Accreditation Bodies and Standardized Medical Record Content

Various accreditation bodies exist that provide accreditation for healthcare organizations. Each have specific requirements that guide standards for documentation. These agencies include the Joint Commission on Accreditation of Healthcare Organizations (JCAHO), the National Committee for Quality Assurance (NCQA), the American Medical Accreditation Program (AMAP), the American Accreditation HealthCare Commission/Utilization Review Accreditation Commission (AAHC/URAC), Det Norske Veritas (DNV), and the Accreditation Association for Ambulatory HealthCare (AAAHC). In addition, the Foundation for Accountability (FACCT) and the Agency for Healthcare Research and Quality (AHRQ) play important roles in ensuring the quality of healthcare (Viswanathan and Salmon 2000).

The Joint Commission on Accreditation of Healthcare Organizations (JCAHO) has two chapters in its accreditation manual that are used in the domain of health information management and information services. They are standards for Record of Care, Treatment and Services and Information Management. These standards outline specific elements of performance that are required to establish effective operating procedures in information management. In some areas, such as orders, they do not specify requirements but rather direct the organization to implement and follow "hospital policies, and medical staff bylaws, rules and regulations (§ 482.24(c)(2) 2016)." As a result, a general standard does not apply but allows for individual adaptation and variation.

The Joint Commission standards have evolved over time and adjusted to accommodate the changing landscape of documentation practices. As paper records have become obsolete, many requirements for documentation corrections or authentication have changed to accommodate electronic systems. For example, the system generated, date, time, and provider signature replace the need for a physical signature in the medical record.

## 3.4 Health Record Content Prescribed by Discipline

Various professional clinical disciplines can have practice standards that impact the manner and method by which and when documentation is captured as part of the healthcare record. These include The American Nurses Association and the American Academy of Physical Medicine and Rehabilitation for example. In addition, hospital governance may direct other standards for documentation practice.

The American Health Information Management Association (AHIMA) is the premier professional affiliation for traditional Health Information Management personnel. They offer advanced certifications in areas of Health Information Administration, Privacy and Security, Documentation Improvement and Data analytics in addition to a variety of medical coding certifications. With more than 103,000 members in 2018, the association continues to have the strongest collective representation in traditional medical records and health information management.

Some states have regulations that require directors of health information departments in acute care organizations to maintain credentials as registered health information administrators (RHIA) or registered health information technicians (RHIT). Both credentials are issued by AHIMA. The credentials provide evidence of core knowledge in industry standards. As electronic health information management continues to evolve, the scope and domain of AHIMA continues to broaden and diversify in order to encompass the many variations in roles and responsibilities that are being adopted by a credentialed AHIMA professional. As the need for data analytics and data integrity specialists continues to expand, more health information professionals may opt for those types of credentials and certifications.

## 3.5 Summation

The health record has developed over time to accommodate the ever-increasing regulatory requirements in healthcare and to provide evidence for medical necessity and treatment for the services provided. As electronic health records have been developed, so has a proliferation and expansion of data sets that support them. Portable Health Records in a Mobile Society will need to be simplified and condensed into core elements that support continuity of care in every potential setting.

## References

§ 482.24(c)(2). JC Record of care, treatment and services. 2016.
Adler-Milstein J, Holmgren AJ, Kralovec P, Worzala C, Searcy T, Patel V. Electronic health record adoption in US hospitals: the emergence of a digital "advanced use" divide. J Am Med Inform Assoc. 2017;24(6):1142–8.

Annas GJ. HIPAA regulations—a new era of medical record privacy? N Engl J Med. 2003;348(15):1486–90.

Blumenthal D, Tavenner M. The "meaningful use" regulation for electronic health records. N Engl J Med. 2010;363(6):501.

Centers for Medicare & Medicaid Services. State operations manual: appendix V – interpretive guidelines – responsibilities of Medicare participating hospitals in emergency cases. 2012. https://www.cms.gov/Regulations-and-Guidance/Legislation/EMTALA/. Accessed 14 Sept 2018.

Gostin LO. National health information privacy regulations under the Health Insurance Portability and Accountability Act. JAMA. 2001;285(23):3015–21.

Spath PL. Role of HIM professionals in quality management. Perspect Health Inf Manag. 2009;6(Summer):1j.

U.S. Government Publishing Office: 45 CFR 164.520. 2002. https://www.gpo.gov/fdsys/search/pagedetails.action?collectionCode=CFR&browsePath=Title+45%2FSubtitle+A%2FSubchapter+C%2FPart+164%2FSubpart+E%2FSection+164.520&granuleId=CFR-2003-title45-vol1-sec164-520&packageId=CFR-2003-title45-vol1&collapse=true&fromBrowse=true&bread=true. Accessed 14 Sept 2018.

Viswanathan HN, Salmon JW. Accrediting organizations and quality improvement. Am J Manag Care. 2000;6(10):1117–30.

# Chapter 4
# The EMR/EHR Marketplace

**Egondu R. Onyejekwe**

**Abstract** The implementation of electronic health records (EHR) is supported by a growing market of EHR providers. Market analysis and reports show an increase in the number of EHR companies competing in the market and greater focus on healthcare informatics. In addition, reports identify healthcare trends and technologies that are very likely to affect the EHR industry in the future.

**Keywords** EHR · Electronic health records · Electronic health record market Electronic health record vendor · EHR vendor · EHR market

## 4.1 Introduction

In this part of the discourse, regarding the market place of electronic medical/health records, electronic health record (EMR) is used interchangeably with electronic health record (EHR). As discussed elsewhere, the electronic medical record (EMR) and electronic health record (EHR) are computer-based patient medical/health record which are at the core of the healthcare delivery system. Over the last three decades, medical institutions and healthcare providers have encouraged the shift towards computerization because it helps them with the management of patient medical/health information. The rush to computerize prevented a well-articulated plan to integrate the information systems. However, by the late 1980s, there were major strides to transform the disparate and individual laboratory computers into integrated clinical information systems, thus allowing for a single terminal extraction of the patient's data such as test results, including blood chemistry, microbiology, radiology, biopsy reports, etc. With time, healthcare providers employed transcription services to incorporated parts of the clinical narrative. Consequently, data sources became integrated in the clinical information systems. Included in this list are sources from surgical operative notes, discharge notes, summary of the

E. R. Onyejekwe (✉)
Public Health, Health Administration, College of Health Sciences, Walden University, Minneapolis, MN, USA
e-mail: Egondu.onyejekwe@mail.waldenu.edu

© Springer Nature Switzerland AG 2019                                   35
E. R. Onyejekwe et al. (eds.), *Portable Health Records in a Mobile Society*,
Health Informatics, https://doi.org/10.1007/978-3-030-19937-1_4

patient's medical problems, and lists of their current medications (Kalorama Information 2018a). The brief history of EMR/EHR therefore sounds the alert that there are different market sizes in this arena. The 2018 Kalorama Information Report includes all aspects of EMR/EHR (Kalorama Information 2018a), and some other estimates focus only on software costs. However, while software costs form major components of competitive revenues in EMR, they do not tell all of story.

Because of the need for currency (as well as the volatility) of the EMR/EHR marketplace, this section focuses mainly on the 2018 Kalorama Information Report market break between physician and hospital EMR/EHR markets. Granted that these could be very different, they also added small hospital and big system revenues, which can yield different results, in their analysis.

## 4.2 EMR/EHR Market Predictions

According the 2017 Kalorama Information Report analysis of electronic health record systems, the size of the electronic health record market was $28 billion in 2016 (Kalorama Information 2017). Their analysis included key profits areas such as revenues for EMR/EHR systems, CPOE systems, and directly related services such as installation, training, servicing, and consulting. It did not include PACS (medical imaging systems) or hardware. The sources of the analysis included vendor reviews, annual reports, interviews with executives, and much more. At that time, they predicted that the EHR market would rise briskly, and Kalorama's forecast is that the market will be $36.6 billion by 2021. But, Kalorama cautioned about investing in the EHR Market, because evolving (see Disruptions below) trends must be offset. Included in their list were: competition, downward pressure on price; and what technology [including emerging technology] would provide the most benefit. Who leads the market will all depend on how these trends play out.

By 2018, however, their analysis for the EMR/EHR market was $29.7 billion in 2017. It is expected to rise to $39.7 billion by 2022 (Kalorama Information 2018b). Included in this their industry annual report of course are: revenues for EMR/EHR systems, CPOE systems, and directly related services such as installation, training, servicing, and consulting which are key profit areas for companies (Kalorama Information 2018b).

This is their Eleventh Edition and consists of 403 pages. It is used by different groups to analyze this EMR/EHR industry. Below is a synopsis of their 403-page report that includes a Global analysis of the EMR/EHR market. The focus is on the market and trends that affect electronic medical/health record software and related services such as "statistics influencing the industry, demographics, life expectancy, and company strategies (Kalorama Information 2018a)." The total market summary includes:

- EMR Market Analysis 2015–2022 ($ millions)
- Revenues and Market Share of EMR Providers 2017 (in millions $)

- Market Size and Growth for Physician/Web-based EMR Market, 2017–2022 ($millions)
- EMR Hospital Market Analysis 2015–2022 ($ millions)
- Hospital vs Physician/Web Breakdown 2017 (%)
- Hospital vs Physician/Web Breakdown 2022 (%) (Kalorama Information 2018a)

The revenues emanate from worldwide figures including: sales of software, services, consulting and replacement. While the market analysis is global in nature, their trend analysis focuses on the U.S. because it is the largest healthcare market and also, the most incentivized for EMR conversion. Their report features EMR Market Analysis for 2017–2022 to include the countries listed below (Kalorama Information 2018a):

- United States
- EMEA
- United Kingdom
- Germany
- France
- Spain
- Nordic Countries
- Middle East
- Israel
- Africa
- Other EMEA
- APAC EMR
- Japan
- China
- India
- Australia
- Other APAC
- Rest of World
- Brazil
- Canada
- Mexico
- Other Rest of World

Below are the different primary issues and trends that affect the electronic medical (health) records (EMR/EHR) industry. These include: demographics, increasing life expectancy, and technology innovation which will continue to fuel growth in the future. There are new developments that will also positively influence growth. So below are the issues and trends explored in their study (Kalorama Information 2018a):

- Patient Engagement
- Information Overload
- Patient Access—Blue Button Technology
- Blockchain

- Healthcare Analytics
- Virtualization Technology
- Interoperability
- Cloud Computing
- Artificial Intelligence and Healthcare
- Big Tech Invasion
- Big Data
- Internet of Health Things
- Healthcare Cybersecurity
- Global Healthcare Spending Trends

EMR 2018 includes is a competitive analysis of leading EMR system providers and the competitors profiled include (Kalorama Information 2018a):

- 4Medica
- AdvanceMD
- Alert Life Sciences Computing S.A.
- Allmeds
- Allscripts Healthcare Solutions
- Amazing Charts
- Aprima Medical Software, Inc.
- athenahealth, Inc.
- Bernoulli Enterprise, Inc.
- BizMatics, Inc.
- Cambio Healthcare Systems
- CareCloud
- Cerner Corporation
- Change Healthcare
- ChartLogic, Inc.
- CompuGroup Medical AG
- Computer Programs and Systems, Inc.
- Computer Sciences Corporation (CSC)
- CureMD
- Dr First/Rcop
- eClinicalWorks
- eMDs
- EMIS Health
- Epic Systems Corporation
- GE Healthcare
- Greenway Health, LLC
- Healthland
- Henry Schein MicroMD
- IBM Healthcare
- IMS MAXIMS
- InterSystems Corporation
- Kareo

- Koninklijke Philips N.V.
- Kronos Incorporated
- McKesson Corporation
- MEDENT Community Computer Service, Inc.
- MEDHOST
- Medical Information Technology, Inc. (MEDITECH)
- Meditab Software, Inc.
- Medsphere Systems Corporation
- Microtest Ltd.
- NantHealth
- NextGen Healthcare Information Systems
- Nextech
- Nightingale Informatix Corporation (now Telus Health)
- NoemaLife S.p.A. (now Dedalus Healthcare Group)
- Nuesoft Technologies, Inc.
- Practice Fusion
- Praxis
- Qualcomm Life, Inc.
- Quality Systems, Inc.
- Quest Diagnostics, Inc.
- SAP SE
- SequelMed
- Streamline MD
- Tieto
- WebPT
- WRS Health

## 4.3   Conclusion

This section concludes with the Healthcare Informatics lists that ranks the first 100 entities. The first 25 entities are listed in full (Healthcare 100 Informatics 2018):

1. Optum, Eden Prairie, MN, $8,087,000,000
2. Cerner Corp., Kansas City, MO, $5,140,000,000
3. Cognizant, Teaneck, NJ, $4,263,405,000
4. Change Healthcare, Nashville, TN, $3,305,100,000
5. Philips, Amsterdam, Netherlands, $3,026,158,434
6. Epic, Verona, WI, $2,700,000,000
7. Dell EMC, Round Rock, TX, $2,350,000,000
8. Conduent, Florham Park, NJ, $1,834,740,000
9. Leidos, Reston, VA, $1,802,000,000
10. Allscripts, Chicago, IL, $1,800,000,000
11. Conifer Health Solutions, Frisco, TX, $1,600,000,000

12. Softheon, Stony Brook, NY, $1,322,441,369
13. Athenahealth, Watertown, MA, $1,220,000,000
14. Wipro Limited, Bangalore, India, $1,200,000,000
15. Tata Consultancy Services, Mumbai, India, $1,000,500,000
16. GE Healthcare, Chalfont St. Giles, U.K., $900,000,000
17. Nuance Communications, Burlington, MA, $899,000,000
18. 3 M Health Information Systems, Murray, UT, $721,000,000
19. Omnicell, Mountain View, CA, $716,200,000
20. Ciox Health, Alpharetta, GA, $631,000,000
21. Wolters Kluwer Health, Waltham, MA, $624,340,000
22. Cotiviti Holdings Inc., Atlanta, GA, $605,228,000
23. Roper Technologies, Inc., Sarasota, FL, $588,000,000
24. Oracle, Redwood Shores, CA
25. IBM, Armonk, NY

# References

Healthcare 100 Informatics. 2018. https://www.healthcare-informatics.com/hci100/2018-hci-100-list. Accessed 11 Sept 2018.

Kalorama Information. The state of the EMR market in 2017. 2017. https://www.kaloramainformation.com/Content/Blog/2017/04/28/The-State-of-the-EMR-Market-in-2017. Accessed 11 Sept 2018.

Kalorama Information. EMR 2018: The market for electronic medical records (physician and hospital EHR market, geographic regions, trends, and issues). 2018a. https://www.kaloramainformation.com/EMR-Electronic-Medical-Records-Physician-Hospital-EHR-Geographic-Regions-Trends-Issues-11633681/. Accessed 11 Sept 2018.

Kalorama Information. EHR-nearing 30 billion in revenue. 2018b. https://www.kaloramainformation.com/Content/Blog/2018/05/21/EHR%2D%2D-Nearing-30-Billion-in-Revenue. Accessed 11 Sept 2018.

# Chapter 5
# US Performance in Healthcare

Egondu R. Onyejekwe

**Abstract** The United States (US) like many advanced countries of the world, is transitioning its healthcare industry from paper to electronic health records (EHRs). During the preceding decade, this move to EHR has marked a critical advance in the US medical and healthcare arena. The US is the leader in the EHR marketplace and continues to grow that area. Unfortunately, the US EHR growth is not commensurate with the provision of care and is a far cry from the cost of healthcare in equally developed countries. What the future would hold for the US healthcare despite current performances, and in the context of electronic (all be it) portable health records is worthy of discourse. The abysmal performance of the US healthcare even in the face of uncontrollable rising costs is of major concern. An emerging area with a promise of solution path is commoditization. Commoditization robs companies of their competitive edge and for that reason, companies strive to avoid it. However, commoditization of healthcare information technology (IT), specifically, the electronic health records (EHRs) seems very compelling if the exploding cost of US healthcare can be curtailed. The drawbacks for commoditization of healthcare IT includes the danger of limiting the advances in the field and that of potential reduction in innovations. The dilemma is that the varied presentations of EHRs need to be portable in order to be useful, and just like software products, meaningful portability requires some standards and modularity. Modularization plus standards are elements of commoditization that would be geared towards a downward spiral of IT costs in the US healthcare field. The jury is still out on what commoditization would yield in the US healthcare industry. Is a "market" system a possible outcome? Would affordable healthcare IT solutions be encouraged? Would commoditization enable hospitals save money on EHRs? These questions remain to be answered as the US seeks solution paths to the uncontrollable costs of its healthcare IT. One potential solution path for sure is that commoditization and standards are enablers for the portability of electronic health records.

E. R. Onyejekwe (✉)
Public Health, Healthcare Administration, College of Health Sciences, Walden University, Minneapolis, MN, USA
e-mail: Egondu.onyejekwe@mail.waldenu.edu

© Springer Nature Switzerland AG 2019                                                            41
E. R. Onyejekwe et al. (eds.), *Portable Health Records in a Mobile Society*,
Health Informatics, https://doi.org/10.1007/978-3-030-19937-1_5

**Keywords** Electronic health records (HER) · Healthcare costs · Commoditization Portability · Healthcare IT

## 5.1 Introduction

As with many advanced countries of the world, the transition of the United States (US) healthcare industry from paper to electronic health records (EHRs) has, during the preceding decade, marked a critical advance in the US medical and healthcare arena. How these advances have affected a nation's performance in the healthcare landscape is worth recognizing. What the future would hold, despite current performances, and in the context of electronic (all be it) portable health records is worthy of discourse. The US Institute of Medicine addresses these six domains of care quality: *safe, effective, patient-centered, timely, efficient, and equitable.* How the US healthcare performs (especially in the era of electronic health records) is paramount.

While there are various ways to measure performance of healthcare in the different countries, we selected areas where there have been substantiated measures that included different metrics. Regardless of measures, what is much more burdensome is the escalating cost of healthcare in the US. It is far above those of other advanced countries. Yet, the quality of care delivered in the US lags behind those in comparatively advanced countries. No viable solution is in sight, but there is hope with the opportunity to commoditize the electronic health records (EHRs). How the US approaches this potential is yet to be determined.

## 5.2 Performance Measures in Healthcare

A comparison of US healthcare to those other developed countries is presented in this Commonwealth Fund publication entitled "Mirror, Mirror 2017: International Comparison Reflects Flaws and Opportunities for Better US Health Care," by Schneider E, Sarnack DO, Squires D, Shah A, Doty MM. Here, the authors (Schneider et al. 2017) compared US healthcare to 10 other high-income developed countries using recent data. They also considered the different approaches to health care organization and the delivery mechanisms that could contribute to top performance. At issue is that the US spends far more on health care system, than the other high-income countries. Spending levels that rose continuously over the past three decades are depicted in Fig. 5.1.

While the increasing and higher cost of healthcare in the US is noted, does the escalating US high cost of healthcare result in quality care? Current studies do not attest nor conform to the norm of the association between healthcare quality and increasing healthcare cost(s). The paper by Sawyer and Gonzales (Sawyer and Gonzalez 2017) applied different charts and metrics for comparing the US

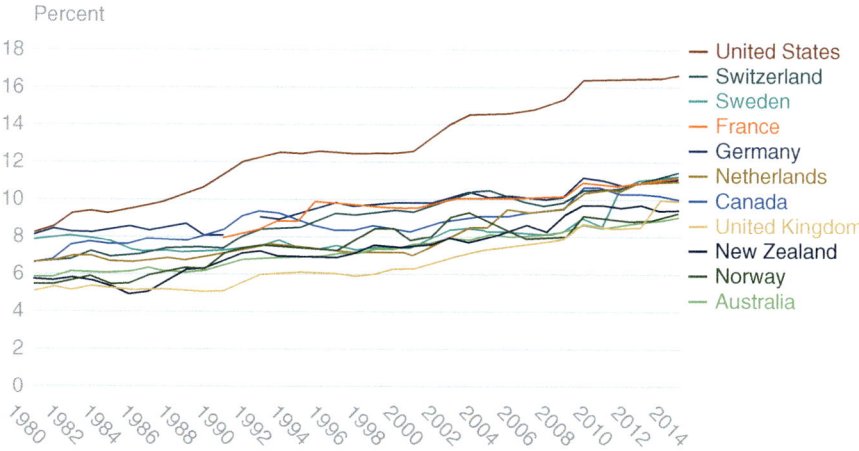

**Fig. 5.1** Healthcare spending as a percentage of GDP, 1980–2014. Reproduced from Schneider et al. (2017)

Healthcare system to other countries in the areas of: health outcomes, quality of care, and access to services. While they documented problems associated with inconsistent or unavailable data and imperfect metrics that made it difficult to accurately judge system-wide health quality in the US, from the data they had, they were still able to make some suggestions. Bench-marking US quality measures against those of similarly large and wealthy countries was one way to assess how successful the US has been. The data they reviewed indicated that the US system is improving across each of these dimensions; however, the data showed that in many aspects, the US continued to lag behind comparably wealthy and sizeable countries at improving care for its population and in learning from systems that often produced better outcomes.

The Organization for Economic Co-operation and Development (OECD) has also compiled data on dozens other outcomes and process measures, across a number of which, the U.S. laged behind other similar OECD countries. (Similar OECD countries are those that are similarly large, and wealthy based on Gross Domestic Product [GDP] and GDP per capita.) As evidenced, mortality rates have seemingly fallen in the US when compared to similar countries. Yet, the fall in the U.S. lags behind those of the other countries. Additionally, the gap seems to be growing between the other countries and the US in areas such as the rates of "all-cause mortality, premature death, death amenable to healthcare, and disease burden (Sawyer and Gonzalez 2017)…"

Furthermore, the U.S. has documented gaps in the quality of care. For example, life expectancy, which improved for decades, succumbed and worsened for some populations in recent years, aggravated by the opioid crisis. Also, as the baby boomer population ages, more people in the US (as with people across the globe) are living with age-related disabilities and chronic disease. These place additional pressures on health care systems to respond. While timely and accessible health care could mitigate many of these challenges, the US healthcare system falls short, because it is ill equipped and fails to deliver indicated services reliably to all those who could benefit. Of particular interest is poor primary care access which has also contributed to a host of inefficiencies. Among them are "inadequate prevention and management of chronic diseases, delayed diagnoses, incomplete adherence to treatments, wasteful overuse of drugs and technologies, and coordination and safety problems (Sawyer and Gonzalez 2017)."

In their 2017 International comparison report, Schneider et al.(Schneider et al. 2017) reflected on both the flaws and the opportunities for better US Healthcare. Health care system performance in Australia, Canada, France, Germany, the Netherlands, New Zealand, Norway, Sweden, Switzerland, the United Kingdom, and the United States were explored using seventy-two indicators selected in five domains. The five domains of interest were:

- Care Process;
- Access;
- Administrative Efficiency;
- Equity; and
- Health Care Outcomes.

Among their data sources were Commonwealth Fund international surveys of patients and physicians and selected measures from OECD, WHO, and the European Observatory on Health Systems and Policies. They then calculated performance scores for every domain and provided an overall score for each country. Here is a summary of their key findings:

As displayed in Fig. 5.2, among the 11 countries in the study, the US ranked last on overall health care systems performance; ranked last in Access, Equity, and Health Care Outcomes; and ranked next to last in Administrative Efficiency, as reported by patients and providers. The US had better performance in Care Process, where it ranked fifth among the 11 countries. Other countries that rank near the bottom on overall performance included France (10th) and Canada (9th).

| header | AUS | CAN | FRA | GER | NETH | NZ | NOR | SWE | SWIZ | UK | US |
|---|---|---|---|---|---|---|---|---|---|---|---|
| **OVERALL RANKING** | **2** | **9** | **10** | **8** | **3** | **4** | **4** | **6** | **6** | **1** | **11** |
| Care Process + | 2 | 6 | 9 | 8 | 4 | 3 | 10 | 11 | 7 | 1 | 5 |
| Access + | 4 | 10 | 9 | 2 | 1 | 7 | 5 | 6 | 8 | 3 | 11 |
| Administrative Efficiency + | 1 | 6 | 11 | 6 | 9 | 2 | 4 | 5 | 8 | 3 | 10 |
| Equity + | 7 | 9 | 10 | 6 | 2 | 8 | 5 | 3 | 4 | 1 | 11 |
| Health Care Outcomes + | 1 | 9 | 5 | 8 | 6 | 7 | 3 | 2 | 4 | 10 | 11 |

**Fig. 5.2** Healthcare system performance rankings. Reproduced from Schneider et al. (2017)

This analysis shows different levels of variations in performance across the indicated domains. There is no country with a perfect score or one that ranks first consistently across all domains or measures. This implies that all countries have room to improve. So, while the US, France, and Canada scored lower than the 11-country average across most of the five domains, all three achieved above-average performance on at least one domain. France scored high on Health Care Outcomes, Canada scored higher on Care Process and Administrative Efficiency, and the US scored high on Care Process.

## 5.3   Top Performers

Overall, there are three top-ranked countries, the United Kingdom (UK), Australia and the Netherlands. The UK in general, and when compared to other countries, achieved superior performance in all areas except Health Care Outcomes, where it ranked 10th (next to last) despite experiencing the fastest reduction in deaths amenable to health care in the past decade. Australia, on the other hand while ranking the highest on Administrative Efficiency and Health Care Outcomes, and while also being among the top-ranked countries on Care Process and Access, it actually ranked low on Equity. Also, whereas, the Netherlands was among the top performers on Care Process, Access, and Equity; its performance on Administrative Efficiency left much to be desired.

As for the others, here is how their performances vary: New Zealand which performed well on measures of Care Process and Administrative Efficiency, performed below the 11-country average on other indicators; Norway and Sweden performed well on Health Care Outcomes compared to the other countries, but had relatively low rankings on Care Process. Finally, Switzerland performed well on measures of Equity and Health Care Outcomes, while Germany achieved a high rank only on measures of Access.

The US health system is an outlier, when based on a wide range of indicators. This is because the US spends far more than the other high-income nations, while simultaneously falling short of the performance of what those countries achieved. The authors surmised that the results suggest that if the US wants to achieve an affordable high-performing health care system that serves all Americans, the US healthcare system should look at other countries' approaches.

Figure 5.3 is a reproduction by the Commonwealth Fund of the trends in amenable mortality for selected countries over the decade spanning 2004 and 2014. The figure was originally produced by the European Observatory on Health Systems and Policies, a partnership that is hosted by the World Health Organization (WHO) Regional Office for Europe. This group supports and promotes evidence-based health policy-making by conducting "comprehensive and rigorous analysis of the dynamics of health care systems in Europe (Schneider et al. 2017)." For both 2004 and 2014, mortality amenable to health care remains very high for the United States.

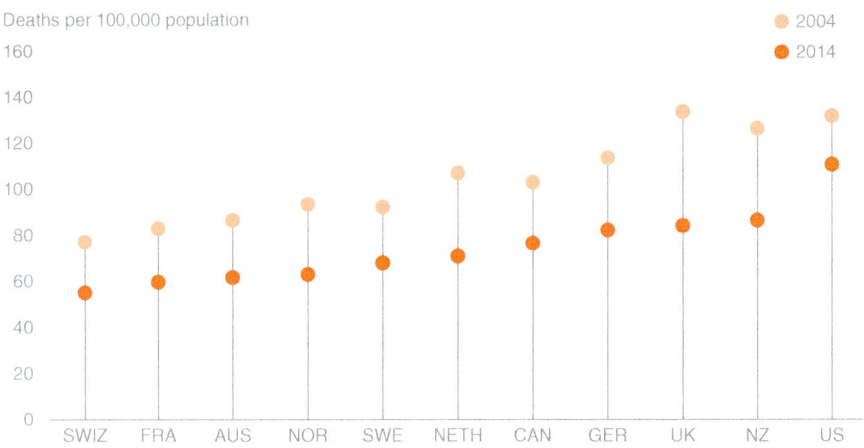

**Fig. 5.3** Mortality amenable to healthcare, 2004–2014. Reproduced from Schneider et al. (2017)

## 5.4   Commoditization: A Solution Path?

What is Commoditization?

"In business literature, *commoditization* is defined as the process by which goods that have economic value and are distinguishable in terms of attributes (uniqueness or brand) end up becoming simple commodities in the eyes of the market or consumers. It is the movement of a market from differentiated to undifferentiated price competition and from monopolistic competition to perfect competition. Hence, the key effect of commoditization is that the pricing power of the manufacturer or brand owner is weakened: when products become more similar from a buyer's point of view, they will tend to buy the cheapest (Wikipedia: Commoditization 2018)."

Is there a role for commoditization in the healthcare industry? In the US as in most advanced countries, the healthcare industry is practically digitized. A cost-effective and sustainable healthcare information system relies invariable on the ability to apply EHR in collecting, processing, and transforming healthcare data into information, knowledge, and action. The many complex and unique problems faced by healthcare providers when implementing such systems, pose several challenges. For all practical purposes, can an EHR from any source become so similar from the consumer's perspective, that they (consumers) will buy what is cheapest in the market?

Indeed, EHRs are everywhere and Big Data has emerged! Furthermore, and in the US *meaningful use*, discussed at length elsewhere, has incentivized the field so much and has resulted in near-ubiquity of electronic health records. Consequently, analytics are also moving towards population, risk-sharing, and value-based care. Also, hospitals and providers have begun to look at what they can do on top of the digitized platform (Siwicki 2017).

"The EHR market has commoditized now and the healthcare domain is coming into an era where most other domains, like the financial domain, have been for a long time—understanding risk, identifying and mitigating risk, and finding tools to do so (Siwicki 2017)," said Fred Rahmanian, chief technology officer at Geneia, a vendor of population health, remote patient monitoring, and analytics systems. "One reason people will see a lot of activity here is because of the ability to ingest a lot of data and extract insights from that data. Healthcare analytics is front and center now (Siwicki 2017)." It is important though for healthcare organizations to understand "the massive troves of data they're sitting on to best function in the burgeoning value-based care market (Siwicki 2017)." The most important thing for health organizations to know in this value-based care and shared-risk environment about the patient population is identifying the risks very early! Healthcare organizations need and must have proper tools that would allow them to properly stratify their populations. Therefore, for healthcare organizations and value-based care, the earlier the provider engages, the better the success rates would be.

Therefore, the argument presented here for commoditization in the current political climate, is that value-based care and risk-sharing models would inevitably emerge to the forefront very soon, because both sides of the healthcare industry (providers and consumers) so desire. Towards that path are emphasis in identifying and mitigating risk at a high level and, even more importantly, reducing the reporting burden on healthcare organizations. However, as this industry moves into risk-sharing models, the reporting requirements would become much more strenuous.

This means that there might be some standard ways to achieve these goals. Invariably, the ways healthcare organizations can identify risk, become the derivatives of the amount of data increases and the evolution of the technology to address them. An example provided here is with regards to the opioid crisis, where it was virtually remote to identify patients at risk for opioid dependency. Now, as troves of data emerge, it may be possible to detect some markers in the collection of medical and prescription claims and identify such patients and/or the dependency. The additional ability to identify new risks would provide additional value.

The discussion about commoditization progresses further with the article titled "Healthcare IT pricing: Can a commoditized future drive down costs (Lichtenwald 2017)?" Commoditization of healthcare takes a different route because in the author's viewpoint, it will not be arriving any time soon.

For people shopping for a laptop, the author opined, it can be an extensive research project where and if the prospective buyer approaches the search with due diligence and compares prices based on various features such as processor speed, monitor resolution, video card quality, and so forth. Such due diligence may not be applicable for a prospective buyer who simply wishes to buy a surge protector because the differences between the variety products are generally so minimal that they tend to focus more on the price. The difference in prices is also true for storage devices, monitors, random-access memory (RAM) etc. where current technological advancement is minimal or incremental. As a matter of fact, and despite the greater complexity, buyers see average prices for laptops to have fallen from the thousands to mere hundreds in recent years. Also, Information technology (IT) services like hosting and internet services are considerably reduced in prices.

Essentially then, commoditization has occurred and is occurring around several IT related areas because the goods and services from the different makers are so similar in design that only their prices differentiate them. "For the consumer, commoditization is good when it brings down prices, but maybe not as good when it represents the slowing of innovation. Obviously, healthcare IT is not currently a commodity in the sense that many IT devices and services are, but the process of commoditization can still have an impact on any industry (Lichtenwald 2017)."

Healthcare IT has two major issues regarding commoditization. First, the current healthcare IT itself is not cheap. Secondly, it is not something for which the average cost can easily be calculated. Take for example an EHR, whose features and functions could pose price tags that range from hundreds of US dollars to billions of U.S dollar, that currently contribute to the rising cost of US healthcare costs. How can the price tag be calculated for it?

So, while a question such as "will commoditization impact healthcare IT and offer some relief for overall costs?" may produce an affirmative answer, it still needs to be properly addressed. Also, will commoditization limit innovation? Indeed, commoditization can hamper innovation after some goals are attained, and this makes it rather urgent to achieving those goals. These potentials are not parts of a near term calculus if only because IT is just not mature enough so it could be standardized in a way meaningful enough to yield results towards competition oriented pricing.

The backlash is that hospitals, for example, cannot operate with too little information about the standard components of a good EHR. They can neither decipher what reasonable prices would look like, nor separate the necessary functions from those which are superfluous or unnecessary. So, the market is imperfect in economic terms given that it is "driven by financial incentives and group think rather than complete information (Lichtenwald 2017)."

In simplistic terms, if commoditization is the goal, then it behooves healthcare IT to achieve commoditization in technology through some agreed-upon modularity and standards. Of course, most standardized hardware is a feature of healthcare IT, but healthcare IT software has no agreed-upon standards. Even within those limits, simply agreeing on a data exchange model and application programming interfaces (APIs) may move the needle closer to some form of uniformity and commoditization. The data exchange model could be the latest HL7 FHIR (Fast Healthcare Interoperability Resources) standard. Without such standards, healthcare information becomes incomplete, resulting in inflated prices, the current feature of healthcare prices in general.

The overall healthcare technology is a big feeder into healthcare costs, while EHRs specifically, in the hospitals, contribute to billions of dollar expenditures. These are systems that arguably are built from healthcare technology that changes progressively each year, but which do not warrant their exorbitant and higher prices per year. This is why, technology change in healthcare is believed to be responsible for at least one third and "as much as two-thirds of per capita health care spending growth (Lichtenwald 2017)."

Can more commoditization in healthcare IT be achieved? Indeed, it would and could be achieved if the range of prices could be narrowed, or if hospitals would better understand massive differences in cost. These would result in higher levels of overall satisfaction with EHRs. This would also enable hospitals to understand what exactly they are paying for, as well as what any system they select would ultimately cost.

Alternatively, commoditization might take a different route. It could take the form of technology deliberately designed as an affordable commodity. The Japanese government, for example, asked Toshiba and Hitachi to design a simpler MRI machine that addressed two basic goals: one that could meet the scanning needs of most patients; and one that is much less than the expensive, than the dominant solutions. By the same token, a technology (the $150 laptop) has been developed by the One Laptop Per Child nonprofit, that introduces children in less wealthy countries to technology.

Other attributes of the healthcare industry in the US are worthy of considerations and resolutions. Among them are: the tension in the interplay between healthcare and costs; US healthcare is run as a business, which requires maximizing shareholder value and profits; but unlike any other business it deals with the health and lives of citizens; the US healthcare is also mostly publicly funded, so decreases in healthcare costs contribute to decreases in the national deficit and debt; and the US healthcare now constitutes 17% of the American economy and continues to rise as the costs continue to spiral (Leonard 2016).

## 5.5  Conclusion

It is noteworthy that commoditization robs companies of their competitive edge, therefore, companies work very hard to avoid it. However, it seems like a recourse and as a matter of fact, a viable path towards curtailing currently unmanageable healthcare costs in the US. But working to avoid it has varied implications on the US healthcare provision side that falls below the standards of the most developed countries of the world, and yet continues to escalate in costs!

On the flip side, the US may never want healthcare IT to fully commoditize "because it would probably result in fewer advances and less innovation." Back to the original premise of this write-up is the portability of EHRs. Because, EHRs are essentially software products, some standards and modularity are essential for porting them. Invariably, such abilities could apply downward pressure on healthcare IT pricing. Nobody knows exactly what a commoditized healthcare IT market would look like, but we'll know some of the principles are having an impact when upper end prices fall to a more reasonable level. Would it even be a "market" system? How would affordable healthcare IT solutions be encouraged? How can hospitals save money on EHRs? While the answers to these questions will usher some solutions, in the US, the disparity between system costs, is still substantial. Many in this US system are hard-pressed to explain the reason/s for the differential costs. To be clear, when commoditization and system standards are factored in, providers, such as hospitals would be able to explain their expenditure and why. Commoditization and standards are enablers for the portability of electronic health records.

# References

Leonard K. US News: US sees historic jump in health care's share of the economy. 2016. https://www.usnews.com/news/articles/2016-12-02/us-sees-historic-jump-in-health-cares-share-of-the-economy. Accessed 5 Aug 2018.

Lichtenwald I. Healthcare IT pricing: can a commoditized future drive down costs? 2017. http://www.medsphere.com/blog/healthcare-it-pricing-can-a-commoditized-future-drive-down-costs. Accessed 5 Aug 2018.

Sawyer B, Gonzalez S. How does the quality of the U.S. healthcare system compare to other countries? 2017. https://www.healthsystemtracker.org/chart-collection/quality-u-s-healthcare-system-compare-countries/#item-start. Accessed 5 Aug 2018.

Schneider E, Sarnack DO, Squires D, Shah A, Doty MM. The Commonwealth Fund: Mirror, mirror 2017: International comparison reflects flaws and opportunities for better U.S. health care. 2017. https://interactives.commonwealthfund.org/2017/july/mirror-mirror/. Accessed 5 Aug 2018.

Siwicki B. EHRs are everywhere: now what? 2017. https://www.healthcareitnews.com/news/ehrs-are-widespread-whats-next. Accessed 5 Aug 2018.

Wikipedia: Commoditization. 2018. https://en.wikipedia.org/wiki/Commoditization. Accessed 5 Aug 2018.

# Part III
# Society

# Chapter 6
# EHR, The Laws and Limits of the Laws

**Egondu R. Onyejekwe**

**Abstract** Electronic health records are subject to several laws and regulations at the federal and state level in the United States. HIPAA privacy and security rules set the floor when it comes to assuring the privacy and security of health information at the national level. While such regulations provide a good structure for single health-care entities or covered entities, they lack flexibility or specificity when it comes to health information environments that may not be proprietary-based. As portable health records evolve, the information will be crowdsourced and managed collectively, which brings up new privacy and security concerns and challenges.

**Keywords** HIPAA · Privacy rule · Security rule · Electronic health records · HITECH Act · Privacy and security · Health information privacy · Health information security

## 6.1 Introduction

Dealing with proprietary data and the vendor hold of Electronic Health Records, in a fragmented marketplace, is further complicated by the law(s). There are essentially, three laws in the United States that relate to healthcare, and specifically to electronic healthcare. They include the Health Insurance Portability and Accountability Act (HIPAA) of 1996 (Wikipedia 2018); The American Reinvestment & Recovery Act (ARRA) that was enacted on February 17, 2009 (HealthIT 2009); and a subset of ARRA—the Health Information Technology for Economic and Clinical Health (HITECH) Act (HealthIT 2009) and others that include many measures to modernize the US infrastructure. The HITECH Act specifically supports the concept of electronic health records—The HITECH Act set meaningful use [EHR-MU], of interoperable EHR adoption in the health care system as a critical national goal (HealthIT 2009). It also incentivized EHR adoption since includes

E. R. Onyejekwe (✉)
Public Health, Health Administration, College of Health Sciences, Walden University, Minneapolis, MN, USA
e-mail: Egondu.onyejekwe@mail.waldenu.edu

© Springer Nature Switzerland AG 2019
E. R. Onyejekwe et al. (eds.), *Portable Health Records in a Mobile Society*, Health Informatics, https://doi.org/10.1007/978-3-030-19937-1_6

both the adoption of EHR as well as the "meaningful use" (HealthIT 2009) of EHR. That is, the use of EHR by providers to achieve significant improvements in care! The effort was led by Centers for Medicare & Medicaid Services (CMS) and the Office of the National Coordinator for Health IT (ONC). This whole idea has been obfuscated by the notion of portable health records which currently are in dispersed and in distributed environments. The implications are discussed later in this chapter.

In any event, this chapter is devoted to the most relevant of the health laws—The landmark piece of legislation in the United States is the Health Insurance Portability and Accountability Act (HIPAA) of 1996. The intent was to "simplify the administration of healthcare, eliminate wastage, prevent healthcare fraud, and ensure that employees could maintain healthcare coverage when between jobs (HIPAA Journal 2018)." The Health Insurance Portability and Accountability Act (HIPAA) of 1996, was enacted by Congress primarily to protect the confidentiality of a person's medical information (Wikipedia 2018). HIPAA therefore, sets boundaries on the use and release of health records as well as, establishes the safeguards to protect the privacy of health information. Despite all its bells and whistles HIPAA addresses the issues required for portable health records (U.S. Department of Health and Human Services 2018).

## 6.2 HIPAA Overview

"The *Health Insurance Portability and Accountability Act of 1996* (*HIPAA*; Pub.L. 104–191, 110 Stat. 1936, enacted August 21, 1996) was enacted by the United States Congress and signed by President Bill Clinton in 1996. It has been known as the Kennedy–Kassebaum Act or Kassebaum–Kennedy Act after two of its leading sponsors (Wikipedia 2018)." The Act consists of five Titles, each of which covers a different topic.

Title I of HIPAA protects health insurance coverage for workers and their families when they change or lose their jobs (Wikipedia 2018). Title II of HIPAA, known as the Administrative Simplification (AS) provisions, requires the establishment of national standards for electronic health care transactions and national identifiers for providers, health insurance plans, and employers (Wikipedia 2018).

Title III sets guidelines for pre-tax medical spending accounts (Wikipedia 2018). Title IV sets guidelines for group health plans (Wikipedia 2018), and Title V governs company-owned life insurance policies (Wikipedia 2018).

HIPAA covers both individuals and organizations and those who must comply with HIPAA are called HIPAA-Covered entities (U.S. Department of Health and Human Services 2018). Covered Entities, Business Associates, and PHI. In general, the protections of the Privacy Rule apply to information held by covered entities and their business associates (U.S. Department of Health and Human Services 2018). According to HIPAA, covered entities are: health plans; clearing houses; a health care provider that conducts certain standard administrative and financial transactions in electronic form; and health associates (U.S. Department of Health and Human Services 2018).

Briefly, HIPAA-covered entities thus, include:

- Health plans: Among these are, Health insurance companies, HMOs, or health maintenance organizations, Employer-sponsored health plans, Government programs that pay for health care, like Medicare, Medicaid, and military and veterans' health programs (U.S. Department of Health and Human Services 2018)
- Clearinghouses, and certain health care providers: Clearinghouses include organizations who on behalf of other organizations, process nonstandard health information to conform to standard data content or format, or vice versa (U.S. Department of Health and Human Services 2018)
- Providers are those who electronically submit HIPAA transactions such as claims. Such providers include, but are not limited to: Doctors; Clinics; Psychologists; Dentists; Chiropractors; Nursing homes and Pharmacies (U.S. Department of Health and Human Services 2018)
- Business Associate—this is a person whom a covered entity engages to help carry out its health care activities and functions. The covered entity must formalize the relationship through a written contract with the business associate or have other arrangement with the business associate that: establishes specifically what the business associate is required to do; and requires the business associate to comply with HIPAA. Included in the business associate's lists are: third-party administrator that assists a health plan with claims processing; consultant that performs utilization reviews for a hospital; health care clearinghouse that translates a claim from a nonstandard format into a standard transaction on behalf of a health care provider and forwards the processed transaction to a payer and an independent medical transcriptionist that provides transcription services to a physician (U.S. Department of Health and Human Services 2018)

Also, a covered health care provider, health plan, or health care clearinghouse can be a business associate of another covered entity (U.S. Department of Health and Human Services 2018).

However, of the five titles, HIPAA Title II is the most relevant to this discourse because it relates to *Privacy* (Wikipedia 2018). The overarching goal of Title II though, is Preventing Healthcare Fraud and Abuse (Wikipedia 2018). Title II contains five rules. The Five Rules of HIPAA Title II include:

1. Privacy Rule
2. Transactions and Code Sets Rule
3. Security Rule
4. Unique Identifiers Rule
5. Enforcement Rule (Wikipedia 2018)

Of these the first—Privacy Rule—and the third—Security Rule—are most relevant for our discussions. The essence of title II of HIPAA which addresses the

Privacy Rule, is to protect most "individually identifiable health information" (U.S. Department of Health and Human Services 2018) that is either held or transmitted by a covered entity or its business associate.

These can be "in any form or medium, whether electronic, on paper, or oral (U.S. Department of Health and Human Services 2018)." Therefore, the Privacy Rule addresses this as Protected Health Information (PHI), which under the US law includes information that can be linked to an individual through any of the following: any information about health status; Information regarding the provision of health care; or information about the payment for health care that is created or collected by a Covered Entity (or a Business Associate of a Covered Entity) (U.S. Department of Health and Human Services 2018).

So, included in the PHI is demographic information, which relates to:

- The individual's past, present, or future physical or mental health or condition
- The provision of health care to the individual or
- The past, present, or future payment for the provision of health care to the individual, and that identifies the individual or for which there is a reasonable basis to believe can be used to identify the individual. Many of the common identifiers of PHI that can be associated with the information above include name, address, birth date, and Social Security Number (U.S. Department of Health and Human Services 2018)

As a consequence, all these—a medical record, laboratory report, or hospital bill—would be PHI because each document would contain a patient's name and/or other identifying information associated with the health data content. Despite all the provisions of the law, HIPAA violations (both privacy and security) are real and common (U.S. Department of Health and Human Services 2018).

By contrast, aggregated data, albeit, compiled from individual health records would not qualify as a PHI. For example, a health plan report that only noted the average age of health plan members as 45 years would not be PHI. Although such a report could have aggregated individual plan member record, no specific individual can be identified.

It is also important to assess the relationship with health information. PHI does not include simply identifying information, such as personal names, residential addresses, or phone numbers. A good example is a phone book, information that is already reported as part of a publicly accessible data source, would not be PHI since it is not related to heath data. Where however, such information would become a PHI is where it was listed with a health condition, health care provision or payment data, with indication that the individual was treated at a certain clinic.

## 6.3   De-identification and Its Rationale

There has been a preponderance of the adoption of health information technologies in the United States to combine large, complex data sets from multiple sources in order to facilitate research and or yield beneficial results. The enactment of the process of de-identification, which enables the removal of identifiers from the health information, mitigates privacy risks to individuals and thereby supports the secondary use of data. These allow for comparative effectiveness studies, policy assessment, life sciences research, and other endeavors!

De-identification of portable health records allows public health professionals and healthcare researchers to conduct epidemiological analysis and clinical investigation with aggregated portable records without compromising the privacy of the participants. The movement to portable records will significantly reduce the size of, or eliminate, consolidated storage of records. These large accumulations of records are used by public health professionals to identify and predict health trends and by researchers to eliminate or control disease. To ensure that these valuable assets are not lost in the march to portable records we must develop standard de-identification processes that can process large numbers of portable records and produce analytical and research databases so that these professionals can continue to work toward improving general health and wellbeing of humankind.

While the Privacy Rule was designed to protect individually identifiable health information through permitting only certain uses and disclosures of PHI provided by the Rule, or as authorized by the individual subject of the information, exceptions are made. One exception is through de-identification (U.S. Department of Health and Human Services 2015a). In recognition of the potential utility of health information even when it is not individually identifiable, §164.502(d) of the Privacy Rule permits a covered entity or its business associate to create information that is not individually identifiable to apply the de-identification standard and implementation specifications in §164.514(a)–(b) (U.S. Department of Health and Human Services 2015a). These provisions allow the entity to use and disclose information that neither identifies nor provides a reasonable basis to identify any particular individual (U.S. Department of Health and Human Services 2015a).

The Privacy Rule provides two de-identification methods:

1. A formal determination by a qualified expert (U.S. Department of Health and Human Services 2015a); or
2. The removal of specific individual identifiers as well as absence of actual knowledge by the covered entity that the remaining information could be used alone or in combination with other information to uniquely identify individuals (U.S. Department of Health and Human Services 2015a). Both methods, even when properly applied, are not foolproof and may yield de-identified data that retains some risk of identification. While the risk is minimal, it is still not zero, the potential for the linking of de-identified data back to the identity of the patient to which it corresponds exists. In any event, and independent of the method, the Privacy Rule does not restrict the use or disclosure of de-identified health information, it is no longer considered protected health information (U.S. Department of Health and Human Services 2015a).

In conclusion, Privacy Rule, while seemingly complex, can be whittled down to these two basics: consent and disclosure. The use of PHI is restricted to six areas: when disclosed to the individual; for treatment, payment and operations; when permission is given; when used incidentally; in benefit of public interest; and when personally-identifiable information has been removed.

## 6.4   Security Rule

Security Rule is best understood as it relates to the Privacy Rule. The difference is that, while the Privacy Rule impacts all forms of PHI, the Security Rule specifically pertains to PHI stored electronically (ePHI) (U.S. Department of Health and Human Services 2018).

### 6.4.1   General Security Rules

The general tenets of the Security Rule require covered entities to apply reasonable and appropriate safeguards for the protection of ePHI. The CMS's Decision tool is useful in determining who the Security Rule covered entities are (U.S. Department of Health and Human Services 2018).

Such entities apply to health plans, health care clearinghouses, and to any health care provider who transmits personally identifiable health information in electronic form. Such transmissions must be in connection with a transaction for which the Secretary of HHS has adopted standards under HIPAA (the "covered entities") and to their business associates (U.S. Department of Health and Human Services 2018).

The safeguards are administrative, technical, and physical. Thus, entities must:

1. "Ensure the confidentiality, integrity, and availability of all e-PHI they create, receive, maintain or transmit;
2. Identify and protect against reasonably anticipated threats to the security or integrity of the information;
3. Protect against reasonably anticipated, impermissible uses or disclosures; and
4. Ensure compliance by their workforce (U.S. Department of Health and Human Services 2018)"

According to the Security Rule, "confidentiality" means that ePHI is not available or disclosed to unauthorized persons and confidentiality requirements "support the Privacy Rule's prohibitions against improper uses and disclosures of PHI [4]." Furthermore, "integrity" under the Security Rule, means that ePHI is not altered or destroyed in an unauthorized manner, while "availability" means that ePHI is accessible and usable when an authorized person needs access to it (U.S. Department of Health and Human Services 2018).

Since HHS recognizes that the range of covered entities span the space between the smallest provider to the largest, multi-state health plan, the Security Rule is flexible and scalable enough to allow covered entities to analyze their own needs and implement solutions that are specific to them and that address their needs. For a covered entity the Rule, rather than dictate measures, requires the covered entity to consider (U.S. Department of Health and Human Services 2018):

- Its size, complexity, and capabilities,
- Its technical, hardware, and software infrastructure,

- The costs of security measures, and
- The likelihood and possible impact of potential risks to ePHI (U.S. Department of Health and Human Services 2018).

Because the healthcare field is continuously changing, it behooves covered entities to also continuously review and modify their security measures protecting ePHI.

The Security Rule also, specifies requirements for safeguards where a HIPAA-covered entity uses ePHI. Those safeguards are broken into three part that include: administrative; physical and technical (U.S. Department of Health and Human Services 2018). Detailed steps are provided for entities in each of these areas.

### 6.4.2  Administrative Safeguards

Administrative safeguards entail a security management process where written privacy procedures are in place to cover authorization, establishment, modification and termination. This implies that a covered entity must not only identify and analyze potential risks to ePHI, but must implement security measures that reduce risks and vulnerabilities to a reasonable and appropriate level. A covered entity must designate a security official who will be responsible for developing and implementing its security policies and procedures. Consistent with the "minimum necessary," aspect of the Privacy Rule, the Security Rule requires a covered entity to only implement policies and procedures for authorizing access to ePHI, when such access is appropriate, based on the user or recipient's role (role-based access). A covered entity must train all workforce members regarding its security policies and procedures, as well as provide for appropriate authorization and supervision of workforce members who work with ePHI. A covered entity must perform a periodic assessment of how its security policies and procedures meet the requirements of the Security Rule. They must also apply appropriate sanctions against workforce members who violate its policies and procedures (U.S. Department of Health and Human Services 2018).

### 6.4.3  Physical Safeguards

Physical safeguards emphasize both facility access and control—where a covered entity must simultaneously limit physical access to its facilities while ensuring authorized access to those thus categorized. Physical safeguards apply also to workstation and device security, where policies and procedures specify proper use of and access to both workstations and electronic media. Overall, physical safeguards require access controls, like security plans, maintenance records and visitor escorts. This includes policies and procedures that govern the transfer, removal, disposal, and re-use of electronic media. These will ensure appropriate protection of ePHI (U.S. Department of Health and Human Services 2018).

### 6.4.4   Technical Safeguards

Technical safeguards focus on access controls that include audit controls; integrity controls; and transmission security. The audit controls include software and or hardware procedures that record and examine access plus other activities in the information systems containing or using ePHI. The integrity controls allow a covered entity to implement policies and procedures that ensure integrity and retention of ePHI. Transmission security safeguards are those controls that guard against unauthorized access to ePHI as it is transmitted over an electronic network. All-in-all, technical safeguards lay out the requirements for use of cryptographic hash functions, data encryption and process documentation. A covered entity must therefore implement technical policies and procedures that would allow only authorized persons access to ePHI (U.S. Department of Health and Human Services 2018).

Time and space do not permit further discussion of other categories of the Security Rule, such as "addressable" and "required" implementation specifications, organizational, policy, procedural, and documentation requirements. Suffice it to state that the essential and pertinent parts of the Security Rule have been addressed. Noteworthy are the key elements of the Security Rule that address who is covered, what information is protected, and what safeguards must ensure appropriate protection and the security of ePHI, including the exporting of such information to other covered entities (U.S. Department of Health and Human Services 2018).

In the piece posted by the 2018 HIPAA Journal, it is hard to not conceive a day without HIPAA violation from either a hospital, health plan, or healthcare professional who is violating HIPAA (HIPAA Journal 2018). There are several and notable updates of HIPAA. They include the HIPAA Privacy Rule, HIPAA Security Rule, HIPAA Omnibus Rule, and the HIPAA Breach Notification Rule (HIPAA Journal 2018). HIPAA is discussed in relation to portable health record and while there are nuances and differences in the updated HIPAA list provided above, they all strive towards the same ends: improving privacy protections for patients and health plan members over the years simply to ensure healthcare data that is safeguarded and the that the privacy of patients is protected. Consequently, a HIPAA violation is a failure to comply with any aspect of HIPAA standards and provisions detailed in 45 CFR Parts 160, 162, and 164 (HIPAA Journal 2018). (These details are not very relevant here, while the violations are relevant.) For those interested in reading more, these are available in the combined text of all HIPAA regulations published by the Department of Health and Human Services Office for Civil Rights, which runs up to 115 pages and contains many provisions.

The concern here is the extraction from the many and hundreds of ways in which HIPAA Rules can be violated. Below is the list of the most common HIPAA violations provided by the 2018 HIPAA Journal (HIPAA Journal 2018):

- Impermissible disclosures of protected health information (PHI)
- Unauthorized accessing of PHI
- Improper disposal of PHI
- Failure to conduct a risk analysis

- Failure to manage risks to the confidentiality, integrity, and availability of PHI
- Failure to implement safeguards to ensure the confidentiality, integrity, and availability of PHI
- Failure to maintain and monitor PHI access logs
- Failure to enter into a HIPAA-compliant business associate agreement with vendors prior to giving access to PHI
- Failure to provide patients with copies of their PHI on request
- Failure to implement access controls to limit who can view PHI
- Failure to terminate access rights to PHI when no longer required
- The disclosure more PHI than is necessary for a particular task to be performed
- Failure to train employees on HIPAA Rules or the failure to provide security awareness training
- Theft of patient records
- Unauthorized release of PHI to individuals not authorized to receive the information
- Sharing of PHI online or via social media without permission
- Mishandling and mismailing PHI
- Texting PHI
- Failure to encrypt PHI or use an alternative, equivalent measure to prevent unauthorized access/disclosure
- Failure to notify an individual (or the Office for Civil Rights) of a security incident involving PHI within 60 days of the discovery of a breach
- Failure to document compliance efforts (HIPAA Journal 2018)

## 6.5  How HIPAA Violations Are Uncovered

Many of these HIPAA violations are discovered though internal audits by HIPAA-covered entities. These can come through supervisors who may have identified employees who have violated HIPAA Rules or directly from employees who often self-report HIPAA violations and potential violations by co-workers. The main enforcer of HIPAA Rules is the HHS' Office for Civil Rights (OCR). Also, it is the OCR that investigates complaints of HIPAA violations reported by healthcare employees, patients, and health plan members as well as investigates all covered entities who report breaches of more than 500 records and conducts investigations into certain smaller breaches. Additionally, OCR intermittently conducts audits of HIPAA covered entities and business associates. Also involved with the investigation of breaches and other HIPAA violations are the State attorneys general especially when reports of breaches of patient records are received (U.S. Department of Health and Human Services 2003; U.S. Department of Health and Human Services 2016; U.S. Department of Health and Human Services 2015b; Andrulis 2010; U.S. Department of Health and Human Services 2007).

### 6.5.1 The Penalties for Violations of HIPAA Rules

The penalties for violations of HIPAA Rules vary—from where State attorneys general can issue high fines and fines that range up to a maximum of $25,000 per violation category, per calendar year; to where OCR can issue fines of up to $1.5 million per violation category, per year. Also, Multi-million-dollar fines can be—as well as have been—issued.

For individuals, there are also potential fines for violating HIPAA Rules and sometimes criminal penalties have been appropriate. Individuals may earn jail terms for violating HIPAA, with some violations carrying a penalty of up to 10 years in jail! Furthermore, healthcare providers, health plans, and business associates of covered entities can also be fined. While more about the penalties for HIPAA violations on this page are available, the 2018 HIPAA Journal (HIPAA Journal 2018) presented the infographics in Figs. 6.1, 6.2, and 6.3 below for a more detailed depiction of recent HIPAA violation penalties and the HIPAA penalty structure.

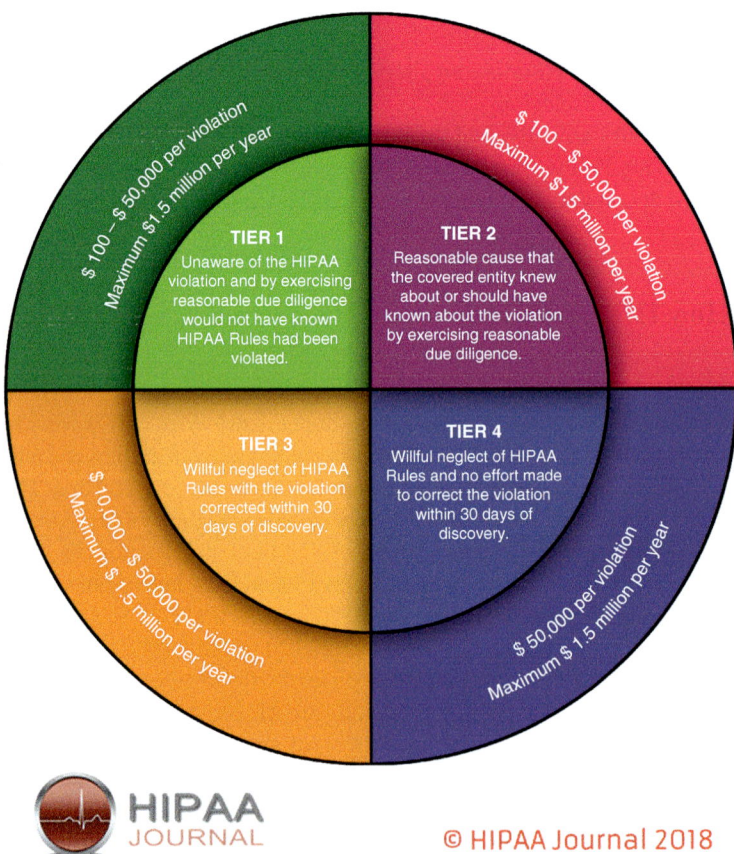

**Fig. 6.1** HIPAA violation penalties. Reproduced from HIPAA Journal (2018)

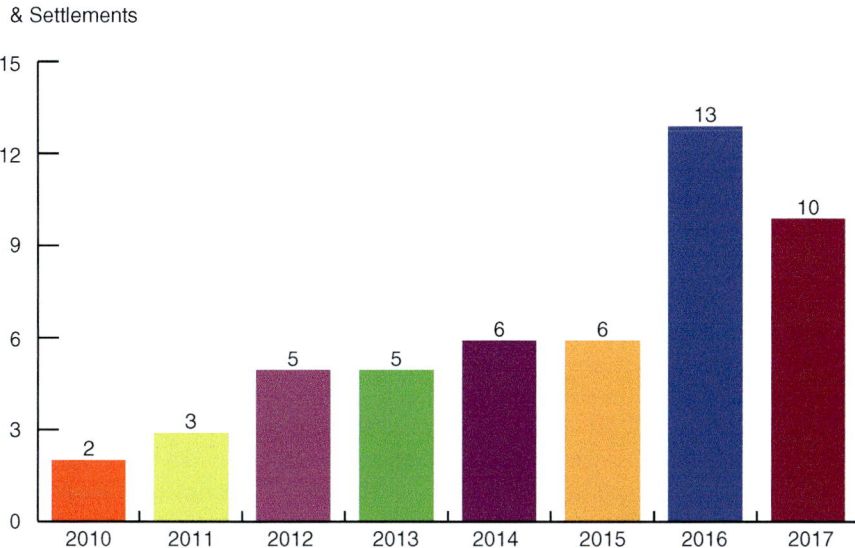

**Fig. 6.2**  HIPAA fines and settlements (2010–2017). Reproduced from HIPAA Journal (2018)

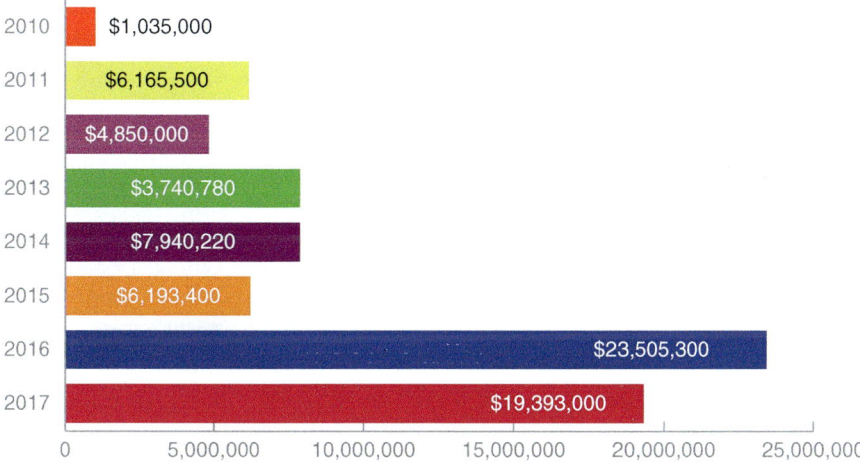

**Fig. 6.3**  Healthcare organizations made to pay for HIPAA violations. Reproduced from HIPAA Journal (2018)

## 6.6   HITECH ACT: Subtitle D—Privacy

### 6.6.1   Part 1: Improved Privacy Provisions and Security Provisions

On November 30, 2009, the regulations provided by Subtitle D, and associated with the enhancements to HIPAA enforcement took effect. These enhancements, although related to privacy, embellished HIPAA.

When data breaches affect 500 or more people, the HITECH Act Subtitle D, requires entities covered by HIPAA to report such breaches to the United States Department of Health and Human Services (HHS), to the news media, and to the people affected by the data breaches. This Subtitle D also extends the liability for the complete Privacy and Security Provisions of HIPAA to the business associates of covered entities, thereby including the extension of updated civil and criminal penalties to business associates. Covered entities are also, required to include these in any business-associate agreements among them (U.S. Department of Health and Human Services 2007).

An additional significant change heralded by Subtitle D of the HITECH Act are new breach notification requirements, that impose new notification requirements on covered entities, business associates, vendors of personal health records (PHR) and related entities if a breach of unsecured protected health information (PHI) occurs. The HITECH Act required both the HHS and the Federal Trade Commission (FTC) to issue regulations associated with the new breach notification requirements. The HHS, on April 27, 2009, issued guidance on how to secure protected health information appropriately. On August 24, 2009, the HHS rule was published in the Federal Register and on August 25, 2009, the FTC rule was published there as well (HealthIT 2009).

A final and significant change made to HIPAA in Subtitle D of the HITECH Act is noteworthy, as it implements new rules for the accounting of disclosures of a patient's health information. The current accounting for disclosure requirements for information used to carry out treatment, payment and health care operations when an organization is using an electronic health record (EHR) became extended. This simultaneously limited the timeframe for the accounting to 3 years instead of its previous 6 years. These changes took effect differentially depending on when organizations were implementing their EHRs (HIPAA/HITECH 2012):

- January 1, 2011, for organizations implementing EHRs between January 1, 2009 and January 1, 2011, and
- January 1, 2013, for organizations who had implemented an EHR prior to January 1, 2009 (HIPAA/HITECH 2012)

HHS, on July 14, 2010, issued a rule that listed categories that included 701,325 entities and 1.5 million business associates who would have access to patient information without patient consent after the patient had given general consent to their medical practitioner's HIPAA release (HIPAA/HITECH 2012).

## 6.7   HIPAA/HITECH Implications for Portability

HIPAA regulations, is the first in the Healthcare industry to provide accepted set of security standards or general requirements for protecting health information. However, and simultaneously, new technologies were evolving, and the healthcare industry began to discontinue processes and rely more heavily on the use of electronic information systems for services such as the payment of claims, answer eligibility questions, provide health information a variety of other administrative and clinically based functions. As indicated earlier, HITECH proposed the *meaningful use of interoperable electronic health records* throughout the United States healthcare delivery system as a critical national goal (Centers for Medicare & Medicaid Services 2010). Meaningful Use itself is defined by the use of certified EHR technology in a meaningful manner (HealthIT 2015). (An example of meaningful manner includes electronic prescribing). Meaningful use also is defined to ensure that the certified EHR technology is connected in a manner that provides for the electronic exchange of health information (HealthIT 2015). Such an exchange should improve the quality of care. Finally, the provider who uses the certified EHR technology must submit to the Secretary of Health & Human Services (HHS) some pertinent information on quality of care and other measures. In summary, the concept of meaningful use rested on the "5 pillars" of health outcomes policy priorities that include (Centers for Disease Control and Prevention 2017):

1. Improving quality, safety, efficiency, and reducing health disparities
2. Engage patients and families in their health
3. Improve care coordination
4. Improve population and public health
5. *Ensure adequate privacy and security protection for personal health information* (Centers for Disease Control and Prevention 2017)

Current Healthcare industry providers, for example, use electronic information systems for clinical applications such as computerized physician order entry (CPOE) systems, electronic health records (EHR), and radiology, pharmacy, and laboratory systems. Health plans provide access to claims and care management, among other applications.

So, portable health records would imply, on the surface, that the medical workforce would be more mobile and more efficient. For example, physicians can check patient records and test results from their location(s). However, a byproduct of the rise in the adoption rate of these technologies is the risk of compromising patient's privacy as well as the potential increases in security risks. A current portable health record does not ensure the adequate privacy and security protection for a person's health record or information because it lies in a distributed environment. This is compromised also because systems so distributed do not allow interoperability of the health records.

This offsets the Security Rule that was designed to protect the privacy of individuals' health information while allowing covered entities to adopt new technologies towards improving the quality and efficiency of patient care. Additionally, the health-care marketplace is diverse, and has great implications for mitigating against both the Security Rule whose design intent allowed for flexibility and scalability. The intent was to allow a covered entity "to implement policies, procedures, and technologies that are appropriate for the entity's particular size, organizational structure, and risks to consumers' e-PHI." Therein lies another problem, as each organization would pursue a route that serves its particular purpose.

## 6.8   Conclusion

True that both HIPAA and HITECH did establish privacy practices for organizations. The nagging questions remain: Can regulations (HIPAA/HITECH) be supported in electronic health record systems that are no longer proprietary-based? In the era of portable health records, the information will be crowdsourced and will (should be) managed collectively. So, it will no longer be managed by entities and or individual organizations. As these individual organizations grow to meet their needs, so grows the problem of privacy and security of electronic health records. Because, the portable health records will, by definition, be crowdsourced, no single custodian (organization) can be held responsible for the security and privacy of the patient record! This means that the (portable) record itself must be inherently secure! Otherwise, access to the records will then be controlled by people who would lack the knowledge and skill needed to maintain the privacy previously required of entities, who were also held accountable to breaches of privacy and security. For health records to be portable, therefore, the implication is that new regulations will have to evolve simultaneously with suitable technologies like Blockchain (distributed ledger). The distributed ledger will support processes like prescriptions to be added to a personal secure health record by any authorized prescribing authority while at the same time maintaining the integrity and the security of the independent and portable health record. At the core remains the management of individual privacy and the security of their health information. Several pertinent questions remain. Who would be responsible for these—privacy and security of patient's portable electronic health record/s? The US Federal government? The state and local governments? Selected and independent organizations? The patient? A patient's agent? Or a mandated electronic health record broker?

## References

Andrulis D. Patient protection and Affordable Care Act of 2010: advancing health equity for racially and ethnically diverse populations. 2010. http://www.jointcenter.org/research/patient-protection-and-affordable-care-act-of-2010-advancing-health-equity-for-racially-and. Accessed 27 Jan 2018.

Centers for Disease Control and Prevention. Meaningful use. 2017. https://www.cdc.gov/ehrmeaningfuluse/introduction.html. Accessed 27 Jan 2018.

Centers for Medicare & Medicaid Services. Secretary Sebelius announces final rules to support meaningful use of electronic health records. 2010. https://www.cms.gov/Newsroom/MediaReleaseDatabase/Press-releases/2010-Press-releases-items/2010-07-13.html. Accessed 27 Jan 2018.

Centers for Medicare & Medicaid Services. Are you a covered entity? 2016. https://www.cms.gov/Regulations-and-Guidance/Administrative-Simplification/HIPAA-ACA/AreYouaCoveredEntity.html. Accessed 27 Jan 2018.

HealthIT. Index for excerpts from the American Recovery and Reinvestment Act (ARRA) of 2009. 2009. https://www.healthit.gov/sites/default/files/hitech_act_excerpt_from_arra_with_index.pdf. Accessed 27 Jan 2018.

HealthIT. Meaningful use definitions and objectives. 2015. https://www.healthit.gov/providers-professionals/meaningful-use-definition-objectives. Accessed 27 Jan 2018.

HealthIT Security. Why HIPAA privacy and HIPAA security rules are needed. 2018. https://healthitsecurity.com/news/why-hipaa-privacy-and-hipaa-security-rules-are-needed. Accessed 27 Jan 2018.

HIPAA Journal. 2018. https://www.hipaajournal.com/. Accessed 27 Jan 2018.

HIPAA/HITECH enforcement action alert. 2012. https://www.morganlewis.com/pubs/eb_lf_hipaaenforcementactionalert_21mar12. Accessed 27 Jan 2018.

U.S. Department of Health and Human Services. Unequal treatment: what healthcare providers need to know about racial and ethnic disparities in healthcare. 2003. http://www.nationalacademies.org/hmd/~/media/Files/Report%20Files/2003/Unequal-Treatment-Confronting-Racial-and-Ethnic-Disparities-in-Health-Care/Disparitieshcproviders8pgFINAL.pdf. Accessed 27 Jan 2018.

U.S. Department of Health and Human Services. The power to reduce health disparities: voices from reach communities. 2007. https://stacks.cdc.gov/view/cdc/12109/. Accessed 27 Jan 2018.

U.S. Department of Health and Human Services. 2015a. https://www.hhs.gov/hipaa/for-professionals/privacy/special-topics/de-identification/index.html Accessed 27 Jan 2018.

U.S. Department of Health and Human Services. HHS action plan to reduce racial and ethnic disparities: implementation progress report 2011–2014. 2015b. http://minorityhealth.hhs.gov/assets/pdf/FINAL_HHS_Action_Plan_Progress_Report_11_2_2015.pdf. Accessed 27 Jan 2018.

U.S. Department of Health and Human Services. National partnership for action to health disparities. 2016. http://www.minorityhealth.hhs.gov/npa/templates/browse.aspx?lvl=1&lvlid=5. Accessed 27 Jan 2018.

U.S. Department of Health and Human Services. Summary of the HIPAA security rule. 2018. https://www.hhs.gov/hipaa/for-professionals/security/laws-regulations/index.html. Accessed 27 Jan 2018.

Wikipedia. Health insurance portability and accountability act. 2018. https://en.wikipedia.org/wiki/Health_Insurance_Portability_and_Accountability_Act. Accessed 27 Jan 2018.

# Chapter 7
# Discrimination

**Dasantila Sherifi**

**Abstract** Discrimination is an issue that still exists in the US despite the protections reinforced by federal laws and education programs provided by private and public organizations. Discrimination based on racial, gender, religious or disability status are observed in employment, education, incarceration, income, and healthcare outcomes. Another form of discrimination is recently created because of the digital divide. Various levels of education, employment, and income, along with lack of portable health records create barriers for patients to access health services and take better charge of their health and health outcomes. Healthcare organizations need to address the various forms of discrimination as they engage in their mission to make health services and health records available to all.

**Keywords** Discrimination · Digital divide · Portable health records · Patient portal access · Healthcare access · Patient engagement

One important aspect of societal factors is the treatment of different categories or groups of people. Federal and state laws protect against discrimination of individuals based on a number of characteristics. Does this mean that there is no discrimination? Research shows discriminatory behavior and practices, as well as discriminatory perceptions still exist and penetrate all the way to healthcare services.

There are three federal laws that address antidiscrimination in employment. Title VII of the Civil Rights Act of 1964 prohibit discrimination on the basis of race, color, sex, religion, or national origin, as well as pregnancy and childbirth related medical conditions; the Age Discrimination in Employment Act (ADEA) prohibits discrimination of employees 40 years and older; and the Americans with Disabilities Act (ADA) prohibits discrimination on the basis of disabilities and requires that employers provide reasonable accommodations for people with disabilities. In addition, a number of states have recently passed laws that protect against discrimina-

D. Sherifi (✉)
College of Health Sciences, DeVry University, Naperville, IL, USA
e-mail: dsherifi@devry.edu

© Springer Nature Switzerland AG 2019
E. R. Onyejekwe et al. (eds.), *Portable Health Records in a Mobile Society*,
Health Informatics, https://doi.org/10.1007/978-3-030-19937-1_7

tion based on the sexual orientation or gender identity and expression, often referred to as lesbians, gays, bisexuals, and transgender (LGBT) antidiscrimination laws. While a number of laws are in place to protect various classes of individuals in terms of employment, direct or indirect discrimination still occurs. A report from US Deputy Permanent Representative to the United Nations (UN) Human Rights Council presented to the UN High Commission for Human Rights on September 14, 2017 showed that the US continues to combat discrimination, stigmatization, and even violence against individuals based on religion (United Nations Human Rights 2017). The report noted a rise in religion-based hate crimes, a rise in workplace religious harassments or lack of accommodation, as well as cases of private employers who denied a sales job or a bus driver's job to Muslim women because of the headscarf they were wearing. The report also showed the ongoing efforts to enforce the laws, train employers, and engage in outreach activities across the country with the goal of building trust and partnerships with the communities. People with disabilities have also experienced discrimination in the workplace in terms of acquiring a job and retaining a job. The ADA Amendment Act of 2008 defined more clearly disability and as a result there was an improvement in job acquisition, job satisfaction, and job retention (Victor et al. 2017). These findings along with ongoing reporting of salary differences among men, women, black men, or Hispanics are examples of evidence that discrimination still exists in our society.

Discrimination is also seen in the higher levels of incarcerations among African American males. Incarceration affects the family life and income during the time of incarceration as well as future employment of the individuals because of their criminal record. This direct discrimination affects individuals and their families in terms of employment and salaries as well as other aspects of their life. Legal protection extended by federal or state laws helps; however, it is not sufficient and negative consequences make their way into inconsistent employment, inconsistent healthcare coverage or healthcare service experiences. When it comes to health records, they may become available to all; however, with inconsistent care providers, there are challenges in obtaining a complete health record. Since health records are not portable, patients in protected classes may not receive the benefit of coordinated and consistent care supported by a historical record of care and physical assessment.

Discrimination based on race or color is felt by black or Hispanic individuals not only in the workplace but also in their encounters with healthcare services. According to Ben, Cormack, Harris, and Paradies, while racism is not associated with the use of health services, it is associated with more negative experiences, including, lower levels of healthcare-related trust, satisfaction, delaying healthcare services and even lack of adherence with the treatment regimen (Ben et al. 2017). Asian immigrant adults who have felt discriminated against experience psychological distress and US-born Asian adults who have felt discriminated because of their ethnic identity experience exacerbated mental health issues (Yip et al. 2008). In a study of microaggressions and discrimination among African Americans, it was found that respondents with darker skin were more likely to be considered as individuals with high-level microaggression types and be classified into the disrespect/condescension category (Keith et al. 2017). When it comes to women, Dehkordy,

Hall, Dalton, and Carlos have found that perceived discrimination was associated with a greater number of visits per year and inverse association with excellent/very good perceived health status (Dehkordy et al. 2016). Basically, women who scored higher on perceived discrimination reported worse health status. Discrimination based on sexual orientation seems to affect older LGBT adults, as well. A study of LGBT and heterosexual individuals' perceptions on long-term care settings found that LGBT respondents were more likely than heterosexual respondents to hide their sexual orientation because they feared the healthcare setting would discriminate against them (Jackson et al. 2017; Sallans 2016).

Healthcare providers may not be discriminating intentionally or even consciously but some patients from certain races or religions certainly feel that way and are more sensitive when it comes to certain conversations that take place. Physicians have all the patient demographics at hand, as well as any other information that is shared through their encounters with patients and their family members. Depending on the implicit bias or perceptions providers create about some patients, they may not start a conversation on a patient portal or portable heath records, which leaves the patient at a disadvantage (especially, if the providers make the wrong assumptions). On the other side, patients' perceptions of being discriminated against lead to lack of trust, lack of sharing the relevant information and an overall gap in communication with the provider, which could lead to patients being less likely to follow up on a provider's note to create a patient portal account, access health information, and overall manage their health matters better by using portable health records.

Discrimination makes its way into education levels, as well. African American, Latino, and American Indian children are generally less prepared to start kindergarten or first grade in terms of language and mathematical skills (Farkas 2012). The gap in education deepens from kindergarten throughout the 12th grade because of the discrimination that occurs by teachers and school administrators. Consequently, fewer individuals from this group complete college education. The gap in education levels may impact their understanding of the importance of accessing their own health information, as well as understanding of medical information contained in their health records. Only 10% of older adults, with limited health literacy, access the internet in order to get health information (Lyles and Sarkar 2015). In general, patients with limited health literacy complete fewer online tasks, take longer to complete online tasks, and have problems in understanding and interpreting medical information accurately (Tieu et al. 2017). As a reminder, the goal of portable health record is not simply for them to become available but to be useful for patients and help patients better manage their health matters. For many individuals that fall into the "less educated" category, making health records accessible on mobile devices may not translate in full access on their end and empowerment in terms of their health issues.

Another way, discrimination may affect access to health records is the digital divide. Multiple studies have found that a digital divide exists between black and non-black patients, when it comes to enrolling, accessing and using patient portals (Ancker et al. 2011; Roblin et al. 2009; Schickedanz et al. 2013; Weingart et al.

2006). Racial differences are also exhibited in the interest people take in discussing or filling a survey about a patient portal. For example, in a study of perceived usefulness and perceived ease of use of patient portals, 95% of the respondents were white. African-American and Asian participation was respectively only 4% and 1% (Sherifi 2018). While this may not be an indication of how people from various races use or perceive the patient portal, it does show racial differences in terms of one or more of these aspects: receiving information from a provider, viewing the information from a provider, prioritizing the follow up action on a provider's message, and taking the time to follow up on a provider's message that is intended to solicit valuable feedback from the patient in the interest of the patient. Lack of regular computer access for certain racial or ethnic minorities is also an issue. Graetz et al. have found that people of certain racial or ethnic minority groups are more likely to access their personal health record exclusively with a mobile device (Graetz et al. 2018). At this time, health records are delivered and made available on the internet by most hospitals and physicians' offices but some may not be fully accessible on mobile devices. In addition, digital divide affects low-income older adults who need assistance with broadband, electronic devices, and internet affordability, as well as subpopulations in the rural areas. As of January 2018, rural Americans were reported to still be at a disadvantage in terms of the broadband internet services (Strover 2018). Many lacked access to internet or had access to broadband services that did not meet the minimum definition of the Federal Communications Commission. Less rural Americans use the internet compared to urban Americans, which means rural Americans are less likely to access their health records online. Wireless internet with decent speed is available but cellphone users need to be in certain geographical proximities to the service towers. Lack of proper access to the internet becomes a barrier to the mobile health records and it also contributes to the computer illiteracy and poor skills in navigating and using web-based applications.

There are other subpopulation that may experience discrimination related to portable health records. Theoretically, people who are ill and need to manage their condition and symptoms on an ongoing basis would benefit greatly by the availability of portable health records. In reality, Cooley et al. found that almost half of the patients undergoing cancer treatment as well as their caregivers struggled to self-manage their symptoms but rarely used computers or mobile devices and identified non-eHealth options for decision support (Cooley et al. 2017). Apparently, being ill or caring for the ill family members seems to be associated with less access to health records and less chances of improving health outcomes.

The various discrimination issues may be hard to address from healthcare organizations and healthcare providers since the system in which they operate is somehow inherently discriminatory; however, there are opportunities for improvements. Just like any problems that need to be addressed, the process should start by acknowledging that a problem exists. Healthcare organizations may already have diversity trainings in place which focus on the value of a diverse workforce, recognizing, preventing, and reporting cases of discrimination or discriminatory harassments. In addition to embracing diversity among colleagues, similar trainings should be expanded to better understand patients or clients (as in the case of health maintenance organizations). Many organizations have started to provide cultural competence trainings for their employees. Learning more about various cultures, religions, or ethnicities helps the healthcare care providers become more sensitive and open minded, possibly better understand patients who may be prone to discrimination, and better understand patient vulnerabilities; all of which can translate into a more genuine, caring, transparent conversation that leads to increased trust between patients and providers.

The science behind identifying the issue of disparities relies on demographic data (Douglas et al. 2015). Identifying disparities through individual studies is great but it is insufficient. There needs to be consistent ongoing documentation, which cannot be done unless the demographic data collection is complete with the data elements that clearly reflect race, ethnicity, disability status, etc. Use of vocabulary standards and consistent categorizations across the continuum of care, as well as federal and state health programs will help with documentation and analysis. EHRs will contribute to better track race and ethnicity data, which should empower providers to understand the disparities better and to address them more effectively. The digital era and the potential that comes with analyzing the data from the various healthcare applications and mobile devices make EHRs "an equity lens" (Rumball-Smith and Bates 2018). EHRs can host additional race-, ethnicity-, or gender-specific data pertaining to treatment preferences or quality indicators. As EHRs are optimized to process all relevant clinical and administrative data, additional opportunities will emerge to not only identify disparities but also intervene and address them.

Another tasks that providers should tackle is the development of menus (within the patient portals) with health information that is reliable, relevant to patient's medical issues, easy to understand, rich with references about human anatomy, medical terminology, and even tools for interpretation in other languages. In addition to health information and health information resources, health technologies must also provide decision support to better help users manage their conditions or healthcare needs. The ultimate goal is to leverage technology for patient education and better management of their own health matters. Furthermore, customized and accessible technology training and support to assist all vulnerable patients and/or caregivers with portal registration and use is also necessary.

From a technological perspective, more can be done to improve access, ease of use, and customization features of patient portals. Better designs would allow for easy downloads and updates of the applications, and most importantly, availability and easy customization of features. For example, it would be more effective for women to be able to access health information or research that relates specifically to women, given the unique aspects of female body, special considerations in medication dosage, and other clinical indicators. LGBT individuals may also appreciate the availability of reliable information that is specific to medical issues which they are experiencing or they need to learn about, while being discreet in obtaining that type of information. Individuals with certain medical conditions, such as cancer or asthma may benefit from use of multimedia and ability to communicate in more than one language through the patient portal. The variety of tools and media available to search for medical information as well as to help in the decision-making process improve the likelihood that people with different backgrounds and different needs will access their health records and engage more in their health matters.

Family practices and primary care physicians may need to play a bigger role in advancing the engagement of all patients with patient-facing technologies and mobile health applications. Since 2014, Baird and Nowak proposed that "primary care providers should become digital health information hubs for their patients" (Baird and Nowak 2014). The current fragmentation of healthcare encounters increases access challenges for health disparate populations. Family or primary doctors are at the center of care and in a position to better coordinate the various services these populations need. As they populate their systems with more granular demographic and clinical information, their chances of having more complete patient records increase. Also, with greater participation in health information exchanges and interoperability, there will be greater possibilities for providers to obtain a more comprehensive patient record. In terms of closing the information loop with patients, more efforts must be dedicated to make such complete, multi-provider, integrated health record available to patients via single sign on portals.

As healthcare organizations embark on a journey to bring health records and health information to consumers, it is imperative to consider discrimination as one of the societal factors that can impact access to and use of health records for better management of health matters. While discrimination may not directly affect access and use of portable health records; there are multiple aspects healthcare providers must consider in order to address discrimination as a barrier in patient engagement through portable health records. Engagement of groups or individuals that are or feel discriminated against requires greater cultural competence and technological offerings. Patient portals and mobile health records may narrow or widen health disparities and often it may seem that discrimination, access to care, access to health records, and quality of care are on a vicious cycle. However, greater, more purposeful, and more optimal use of health IT and mobile devices bring more hope and promise in addressing and eliminating healthcare disparities.

# References

Ancker J, Barrón Y, Rockoff M, Hauser D, Pichardo M, Szerencsy A, Calman N. Use of an electronic patient portal among disadvantaged populations. J Gen Intern Med. 2011;26(10):1117–23. https://doi.org/10.1007/s11606-011-1749-y.

Baird A, Nowak S. Why primary care practices should become digital health information hubs for their patients. BMC Fam Pract. 2014;15:190. https://doi.org/10.1186/s12875-014-0190-9.

Ben J, Cormack D, Harris R, Paradies Y. Racism and health service utilization: a systematic review and meta-analysis. PLoS One. 2017;12(12):e0189900. https://doi.org/10.1371/journal.pone.0189900.

Cooley ME, Nayak MM, Abrahm JL, Braun IM, Rabin MS, Brzozowski J, Berry DL. Patient and caregiver perspectives on decision support for symptom and quality of life management during cancer treatment: implications for eHealth. Psycho-Oncology. 2017;26(8):1105–12. https://doi.org/10.1002/pon.4442.

Dehkordy SF, Hall KS, Dalton VK, Carlos RC. The link between everyday discrimination, health-care utilization, and health status among a national sample of women. J Women's Health. 2016;25(10):1044–51. https://doi.org/10.1089/jwh.2015.5522.

Douglas MM, Dawes DE, Holden KB, Mack D. Missed policy opportunities to advance health equity by recording demographic data in electronic health records. Am J Public Health. 2015;105:S380–8. https://doi.org/10.2105/AJPH.2014.302384.

Farkas G. Racial disparities and discrimination in education: what do we know, how do we know it, and what do we need to know. 2012. https://brainmass.com/file/1474172/Racial+Disparities +and+Discrimination+in+Education-+What+Do+We+know,+How+Do+We+Know+It,+and+ What+Do+We+Need+to+Know-.pdf. Accessed 16 Feb 2018.

Graetz I, Huang J, Brand RJ, Hsu J, Yamin CK, Reed ME. Bridging the digital divide: mobile access to personal health records among patients with diabetes. Am J Manag Care. 2018;24(1):43–8. http://www.ajmc.com/journals/issue/2018/2018-vol24-n1/bridging-the-digital-divide-mobile-access-to-personal-health-records-among-patients-with-diabetes. Accessed 16 Feb 2018.

Jackson NC, Johnson MJ, Roberts R. The potential impact of discrimination fears of older gays, lesbians, bisexuals and transgender individuals living in small- to moderate-sized cities on long-term health care. J Homosex. 2017;54(3):325–39. http://www.tandfonline.com/doi/abs/10.1080/00918360801982298. Accessed 16 Feb 2018.

Keith VM, Nguyen AW, Taylor RJ, Mouzon DM, Chatters LM. Microaggressions, discrimination, and phenotype among African Americans: a latent class analysis of the impact of skin tone and BMI. Sociol Inq. 2017;87(2):233–55. https://doi.org/10.1111/soin.12168.

Lyles CR, Sarkar U. Health literacy, vulnerable patients, and health information technology use: where do we go from here? J Gen Intern Med. 2015;30(3):271–2. http://dx.doi.org.ezp.walde-nulibrary.org/10.1007/s11606-014-3166-5. Accessed 16 Feb 2018.

Roblin DW, Houston TK, Allison JJ, Joski PJ, Becker ER. Disparities in use of a personal health record in a managed care organization. JAMIA. 2009;16(5):683–9. https://doi.org/10.1197/jamia.M3169. Accessed 16 Feb 2018.

Rumball-Smith J, Bates DW. The electronic health record and health IT to decrease racial/ethnic disparities in healthcare. J Health Care Poor Underserved. 2018;29(1):58–62.

Sallans RK. Lessons from a transgender patient for health care professionals. AMA J Ethics. 2016;18(11):1139–46. https://doi.org/10.1001/journalofethics.2016.18.11.mnar1-1611.

Schickedanz A, Huang D, Lopez A, Cheung E, Lyles C, Bodenheimer T, Sarkar U. Access, interest, and attitudes toward electronic communication for health care among patients in the medical safety net. J Gen Intern Med. 2013;28(7):914–20. https://doi.org/10.1007/s11606-012-2329-5.

Sherifi D. Perceived usefulness and perceived ease of use impact on patient portal use (order no. 10744298). Available from ProQuest Dissertations & Theses Global. (2009113067). 2018. http://ezp.waldenulibrary.org/login?url=https://search-proquest-com.ezp.waldenulibrary.org/docview/2009113067?accountid=14872. Accessed 16 Feb 2018.

Strover S. Reaching rural America with broadband internet service. Chicago Tribune. 2018 Jan. http://www.chicagotribune.com/sns-reaching-rural-america-with-broadband-internet-service-82488-20180117-story.html. Accessed 16 Feb 2018.

Tieu L, Schillinger D, Sarkar U, Hoskote M, Hahn KJ, Ratanawongsa N, Ralston JD, Lyles CR. Online patient websites for electronic health record access among vulnerable populations: portals to nowhere? J Am Med Inform Assoc. 2017;24(e1):e47–54. https://academic-oup-com.ezp.waldenulibrary.org/jamia/article/24/e1/e47/2631487. Accessed 16 Feb 2018.

United Nations Human Rights. Office of the High Commissioner for Human Rights: antidis-crimination library. 2017. https://adsdatabase.ohchr.org/IssueLibrary/Forms/ByCategory3. aspx?&&View={fc86de9e-b67d-4384-8d83-a721d4415065}&RootFolder=%2FIssueLibrary &FilterField1=Issues&FilterValue1=278%2C171%2C233%2C180%2C232%2C234%2C23 8%2C332%2C197%2C199%2C205%2C230%2C221%2C229%2C266%2C235%2C236%2 C239%2C253%2C255%2C308&FilterLookupId1=1&FilterOp1=In&TreeField=Issues&Tr eeValue=0c3fd3de-f298-4567-b058-ddd8de18c176&OverrideScope=RecursiveAll&Process QStringToCAML=1&SortField=Date&SortDir=Desc#ServerFilter=FilterField1=Issues-Filt-erValue1=278%2C171%2C233%2C180%2C232%2C234%2C238%2C332%2C197%2C199

%2C205%2C230%2C221%2C229%2C266%2C235%2C236%2C239%2C253%2C255%2C 308-FilterLookupId1=1-FilterOp1=In-TreeField=Issues-TreeValue=0c3fd3de%2Df298%2D4 567%2Db058%2Dddd8de18c176-OverrideScope=RecursiveAll-ProcessQStringToCAML=1. Accessed 16 Feb 2018.

Victor CM, Thacker LR, Gary KW, Pawluk DV, Copolillo A. Workplace discrimination and visual impairment: a comparison of equal employment opportunity commission charges and resolutions under the Americans with Disabilities Act and Americans with Disabilities Amendments Act. J Vis Impair Blind. 2017;111(5):475–82.

Weingart SN, Rind D, Tofias Z, Sands DZ. Who uses the patient internet portal? The patient site experience. JAMIA. 2006;13(1):91–5. https://doi.org/10.1197/jamia.M1833. Accessed 16 Feb 2018.

Yip T, Gee GC, Takeuchi DT. Racial discrimination and psychological distress: the impact of ethnic identity and age among immigrant and United States-born Asian adults. Dev Psychol. 2008;44(3):787–800. https://doi.org/10.1037/0012-1649.44.3.787. Accessed 16 Feb 2018.

# Chapter 8
# The As

**Dorcas Waithira Maina and Dasantila Sherifi**

**Abstract** Access to health records is fundamental in the promotion of universal access to health care. Access reflects the fit between the characteristics and expectations of the health care providers and the clients. Portable health records promote this by ensuring that health information is available where and when it is needed to facilitate care. There are five dimensions of access to health records: availability, affordability, acceptability, accommodation, and accessibility. These components for a chain that is no stronger than its weakest link. The records must be available at the point of care in a manner that facilitates their usage without suffering financial hardship and meeting the preference of the client. In addition, there is emphasis on the characteristics of the health care provider that may influence acceptance of the records. The 5As address the access barriers in adoption and use of portable health records. This chapter addresses these factors in the context of diverse health care settings.

**Keywords** Accessibility · Availability · Affordability · Acceptability · Accommodation · Portable health records

## 8.1 Introduction

Access to quality health care is a universal need. When patients receive care in health facilities, records are generated. The type and form of records depend on the healthcare organization, technological and personal factors, and applicable laws, policies, and regulations. The rationale behind the use of portable health records is

D. W. Maina (✉)
Department of Medical Surgical Nursing, University of Nairobi, Nairobi, Kenya
e-mail: dorcas.maina@waldenu.edu

D. Sherifi
College of Health Sciences, DeVry University, Naperville, IL, USA
e-mail: dsherifi@devry.edu

© Springer Nature Switzerland AG 2019
E. R. Onyejekwe et al. (eds.), *Portable Health Records in a Mobile Society*,
Health Informatics, https://doi.org/10.1007/978-3-030-19937-1_8

to make them readily available at the point of care. This partly stems from the need for comprehensive health information at the point of care and need for quality care. Essentially, if the patients' health information is readily available, it can be used during routine care as well as during emergencies. This chapter will describe how portable health records improve availability, access, affordability, acceptability, and accommodation of care.

## 8.2   Portable Health Records and Availability of Care

Traditionally, patient health records have been paper based and confined to the custody of a single health provider. For these records to be accessed, the patient had to be attended to at that particular facility. If another provider or hospital needed access to the records, they had to request them, which would take time and possibly delay care.

Looking back, the vaccination card (sometimes referred to as an immunization card) that shows a child's birth history and developmental stages has been in use for a long time (Detmer et al. 2008). This is a good example of a portable health record. The parent/guardian, acting as the child's proxy, carries the record to the health care facility where it is updated, and subsequently presented at every visit. This record contains the health history of the child from birth. While it may have been digitized by most physicians' offices, this record is still in use in paper form in some settings. In fact, many parents can retrieve this record upon request, as it may be required for certain jobs, travel abroad, or school registration.

Other forms of portable health records are the medical alert bracelets, allergy tags, and blood type bracelet. More recently, though not well accepted, the use of medical alert tattoos placed in a prominent place has emerged. It is common practice today for health care providers to check the wrist or neck of patients for these records, in addition to questioning them. Their importance is already known especially when caring for the unconscious patient who cannot provide health data.

With advancements in technology, patient's health information is now being stored electronically making access to the data in all situations easier. Further, patients have the option to generate and handle their own health electronic records from various hospitals and outpatient providers (Cushman et al. 2010). Availability of this information in a timely manner can have significant effects on the provision and outcome of care.

According to the Merriam-Webster dictionary, availability refers to the quality of being able to be used or obtained (Merriam-Webster n.d.). This means easily acquired or possessed. When applied to portable health records it refers to the quality of patient's health information being readily obtained when needed.

With portable health records, patients are encouraged to generate their own records in a format that addresses their needs. Using telecommunication devices like tablets, mobile phones, and personal digital assistants, the patient can ensure availability of their records in a mobile format (George and Hopla 2015). While they are recommended for all, these records are especially important for patients who have ongoing health conditions, and whose interactions with health providers produce numerous data. Given the large volume of data, it may be easy to forget the information from different providers. Lack of pertinent information at the point of care can contribute to medical errors (Han et al. 2016). Portable health records have the potential to help the patient and the provider identify health conditions better.

From the perspective of the doctor, availability of information in a timely manner has implications on the outcomes of care. For example, when a patient cannot remember all the prescription drugs they are on, the doctor may prescribe drugs that interact with the ones they are already taking. In the presence of relevant health information, clinical decision-making is based on a thorough health history as well as the physical examination (Radhakrishna et al. 2014). Though the records may be summarized, they provide snapshots of the patient's health status. Coupled with an interview, they can provide pertinent information for care. As a result, no patient is turned away based on lack of health information, neither do they have to present for care to a particular provider. Health care access hence becomes convenient.

During emergencies, the patient is attended by the nearest health provider, while his or her records may be under the custody of another provider. Any delays in such situations can be tragic. Unfortunately, relevant medical information is needed even more urgently for patients in emergencies. The absence of health data means that the health care providers would be cautious while taking care of the 'unknown' patient. They do not have the benefit of knowing relevant health information about the patient in the absence of health records or an informant who can provide it. Further, in such situations, within a short period of time, the patient is handled by multiple providers whose main goal is to preserve life. Portability of health records ensures availability at the point of care, thereby improving and accelerating clinical decision-making, and consequently the outcomes of care. Assuming the patient has the records and that information on allergies, medication lists, and other diagnoses can be retrieved, and these obviously influence the type of care given to the patient.

In other situations, if the patient is at home in an emergency, a friend or relative can provide the records on their behalf (by proxy). This concept of access via proxy also affects adolescents and children. Proxies access information on behalf of the patient. However, when dealing with adolescent with special circumstances like sexually transmitted infections (STIs) or pregnancy, then the applicability of portable health records may be at risk. Though the access to these records by the proxy must get authentication and consent by the patient, certain laws may bar the proxy from accessing such records. In such cases, Hlamka, Mandl, and Tang propose that standards and policies should be put in place to address the needs of this population (Hlamka et al. 2008). When available at the point of care, the portable health records:

- Improve clinical decision making;
- Enhance quick admissions due to efficient health interviews;
- Guide quick action in emergencies situations;
- Promote continuity of care during travel.

Portable health information is also helpful in the event of illness during travel. In these far places, where in some situations the care may be below or above pars to what one is used to, the availability of health information becomes important. In some states or countries, interoperability of health records is not possible due to patient's privacy (Cushman et al. 2010). Laws may prohibit the custodian of the records from sending health records to a party across the border. However, with portable of poor records, the patient chooses who accesses the records. Through prior plans, the patient can request for the records from the custodian, summarize them, and carry the records as they travel. The focus would be to have their health information available in case of any eventualities.

In some countries, medical tourism is now being recognized as a national industry. More and more people would be willing to cross borders if they are assured of quality medical care at a lesser cost. For these patients, portable records that may not be restricted by privacy laws play a major part in promoting care provision.

## 8.3   Portable Health Records and Access to Care

Access to personal records is not a new concept. Although in the USA the Health Insurance Portability and Accountability Act (HIPPA) grants patients access to their own records, few patients take advantage of this (Peacock et al. 2017). In Canada, the patient's information is protected at the federal and provincial levels. Equally, the National Health Service (NHS) has patient data protection laws. To make it even easier, they have opportunities to access their records using available technologies. In Kenya, access to medical records is protected by privacy laws. Though these laws address the rights of the patient to access their health records, the rights of proxies are less clear. However, a similar right should be assumed especially, when access to such records is vital in making health care decisions for the patient.

Access to care is facilitated by presence of relevant information. Of course, it is assumed that the information is standardized to enable interoperability. Some patients may decline to seek care from other health care providers simply because these providers do not have their complete health records. They may feel that the usual provider understands them better; since they have all the information; and the patient does not have to repeat the health history, as would be the case with a new provider.

Comparability of health records is important in ensuring continuity of care. When the patient has ongoing access to the records, it makes it easier to retrieve information for comparison purposes. This retrieval would otherwise be done from the hospital's system or requested from the primary provider. This is associated with delays in access to care because of the retrieval time. These delays can be overcome by portability of records. Though they may hold less information compared to the original health records, they can provide an overview of the patient's health history, thereby saving time.

The consumer carries the portable health record with them across and within the health care system. With a proper design, these records may enhance communication between the patient and the caregiver. In fact, they can be considered as health at hand. By using them, the health care provider can use the opportunity to educate the patient on personal health matters. This encourages greater participation in their care because he or she understands the benefits of accessing their own health information (Delbanco et al. 2012).

Another potential gain for the patient resulting from the use of portable records is that since access to the records is facilitated, the health provider has more time with the patient. By making retrieval of health information less labor intensive, the physician can spend more time with the patient and offer personalized health care.

Some apps have embedded tools, which allow the patient to assess and monitor the physician's accessibility (Palen et al. 2012). Consequently, they can collect reports when the physician is available. They also allows patients to share information electronically, further enhancing the hospital experience by proactively providing all the information in the right sequence.

Nonetheless, the real question is, are patients interested in accessing their health records? Is it not too much work? Though encouraging portability of records ensures that they are easily accessible at the point of care, the patients must understand the importance of these records in accessing care. However, it must be understood that some patients may view access to their records as too much work and may be anxious about the information therein. Some studies have shown that the patient who accessed their records felt reassured because they could confirm that things were alright, some felt that the doctor may not be hiding anything, and others felt reassured that they can access their data at will (Ralson et al. 2004).

While the assumption is that health records are easily accessible, there is a concern that not all patients are literate enough to understand and use them. Health literacy can influence the use of the records since it translates to low self-efficacy. Weitzman, Kaci, and Mandl found that low technological literacy could affect acceptability of portable health records (Weitzman et al. 2009). Assessment of health literacy is important, and the records should be customized to meet the needs of these patients. This may be a difficult and an expensive endeavor. Therefore, where acceptability is slow, the use of these records should be prioritized to target patients with greater health care needs, such as patients with multiple chronic conditions. Further, the cost, privacy, and benefits of these records should be considered.

## 8.4   Affordability of Care by Use of Portable Health Records

Health care is as unique as the patients themselves are. They rarely receive health care from only one organization. In fact, they attend care at different locations like the consultation room, pharmacy, and radiology. Information generated from the various providers contributes to medical decision-making. Every day, these patients make decisions regarding their care, which involves a tradeoff; giving up something in order to gain another, and choosing the decision that means more to them. Getzen noted that people often make choices that make them better off and not necessarily gain financially (Getzen 2013). The choice made in health care is about quality rather than monetary gain.

Maintaining a portable health record is considered an investment in health. The purpose of such records is not simply the immediate use but the long-term improvements of health. The hope is that the patient is willing to invest in time and effort to be prepared for any future health issues.

As it is, the patient may be struggling to pay for basic health care. In order for portable health records to remain meaningful, they must be carried in a standardized format that is easily accessible and updated. Format may vary from simple storage devices to those with added functionalities allowing auto population and interoperability. This means the data could be saved on cloud, in a flash drive, in a tablet, or in the phone.

In resource poor settings, portability of records may be hampered by the use of tools that require maintenance and updates. Since few may afford such tools, it is preferable that portability be enhanced by using cost effective and easily accessed methods. In some cases, the records need not be in an electronic format; paper records would suffice. The only expectation is that all the patients data is aggregated and in a standardized format.

Who then should bear the cost? In addition, is it worth investing in portable health records? The answer depends on the patient's situation and motivation. As it is, several stakeholders are involved in the use of the health records; patients, health care providers, policy makers/regulators, and governments. The issue of who pays for maintenance of these records then arises. For example in developing countries, most patients may choose paying for care versus maintaining portable records. Such patients may not afford fees associated with downloading or updating apps that promote portability. Though Spil and Klein feel that patients should pay since they are the direct beneficiaries (Spil and Klein 2014), it is argued that patients will not do so unless records are relevant to their needs at that time (Kerns et al. 2013). In resource poor settings, the patient must feel that portability of health records is both relevant and cost effective. To promote portability in these settings, simple applications can be used to summarize health history, thus offering a cheaper alternative that still meets the intended use.

There have been cases where certain tests had to be repeated because the provider could not ascertain if they had been done, or they do not have the results. This translates to more costs for the patients. In fact, the patient may know the tests were done and know the results but cannot remember them or report them accurately. In such a case, to avoid errors, the physician may choose to repeat the tests. Such tests, can be avoided when the portable health records are available.

## 8.5    Portability of Records and Acceptability of Care

With the introduction of any new ideas, there are early accepters and late accepters. With all the advancements, the need for data interchange and integration remains integral. In some settings and among certain populations, there is anxiety with adoption of technology. Although all patients can benefit from the use of electronic records, the elderly patients who have more health care needs tend to hesitate to adopt any new technology especially if the benefits are not clear (Price et al. 2013).

While the maintenance of portable health records is not shifted to the patient, the patient still needs to know how to access and present them at the point of care. The expectation is that patients and providers would readily accept and adopt electronic records. However, that is not always the case. For example, in the Kenyan setting, the use of these records may face opposition from the health care providers. Generally, the health care provider is the custodian of the health records, and may feel they are best positioned to do so. In fact, some believe that patients may make changes to the original record. This skepticism is not unfounded and has to be addressed if acceptance of these records in similar resource poor setting is expected. Kerns asserts that though the record can revolutionize care and improve communication, the health care provider may feel like their responsibility over the record is terminated (Kerns et al. 2013). While this is not necessarily the case, it may result to some providers reporting less information in the record since they feel they are under scrutiny (Yau et al. 2011). Alternatively, they may record only what they feel is necessary for certain situations or use medical jargon that is not understood by the patient. Though this may be unique to the Kenyan setting or similar environments, this should not deter the use of portable health records. There are in fact some notable uses with respect to the immunization card, the tuberculosis (TB) drug card, and the antiretroviral drug record among others. In the process of maintaining these records, the provider collaborates with the patient who clearly understands what is contained in the record. The success of these records means that portability of records is possible and should be encouraged.

The one major drawback to acceptability of the records is the lack of consensus on whether patients should access these records. As earlier alluded to, use of portable records promotes speedy access to care. However, physicians may not be united on whether or not the patient should access their own data in the Kenyan setting. Arguably, anticipation of confusion over health terms, medication dosages, test interpretations, and results may be the underlying reason (Lester et al. 2016). Still, these records not only help the patient track their record and appreciate the rationale for decisions made, but also help them play an active role in their health management. When used among patients with coexisting conditions, the coordination of care is made easier and duplication of care is avoided.

Before allowing patients to take charge of their health records, it is important to consider whether they are interested in seeing their records in the first place. Patients may understand that they have the right to access their records but they also have anxiety especially over concerns of confidentiality and privacy (Vodicka et al. 2013; Kahn et al. 2009).

In the current environment where patients easily access information over the internet, even in resource poor settings, access to health records can be promoted. By doing so, the patients feel they are in control of their care. Though the concern over the understanding of the content of these records still looms, the patient can in fact act as quality controllers. In the Kenyan setting, the health care system is burdened by many patients who have to be attended to on a daily basis. The physician still has to keep a record of the encounter with the patient. Through their discussion, the patient and the physician discuss the approach to care which is documented in their record. If there are any errors or any omissions, the patient can easily point them out. To improve acceptability, patients should be encouraged to embrace these records. However, the systems should be in place to help them understand their records, and personalize them to their needs, wishes, and comprehension ability (Lester et al. 2016).

The portability of health records has implications on the physicians and other health care providers. Adoption and acceptability of these records may have implications on their workload. In resource poor settings, the health providers work with limited resources to attend to a large number of patients. Portability of records and collaborative care may somewhat ease this burden. When patients start to understand and interpret their health records, they are better informed about their care. What implications does this have on the physicians? While some think that patients may contact them more for clarification, arguably these records may actually decrease the workload especially if the patient is able to contact the physician online (Ross et al. 2005). This implies that they can get advice from the physicians virtually. In fact, there is an enhanced level of communication because the patient can ask questions that are more relevant to their care. Further, the patient can better understand the information given by the physicians and the patient is able to appreciate the reasons for the treatment. Overall, it promotes trust, confidence, and positive outcomes.

With the perceived issues of acceptability, how do health care organizations respond to patients who have their own records? First, portable health records serve as a bridge to a different environment (Kraan et al. 2015). Normally, in the Kenyan setting, even with hospital-based records the clinicians reviews the previous records of the patient. The only drawback is that there are no designated hospitals that the patient attends. In fact, a detailed record may only be found if the patient has ever been admitted in the hospital. This means that every time the patient attends a health facility even for treatment of minor illness the health provider must obtain fresh data in order to care for the patient. In such case, they rely on the patient's memory of prior care. This can be circumnavigated by ensuring that the most relevant and core information about the patient is available at the point of care by the portable health record. This enhances efficiency of health communication and exchange. In fact, when health information exchange is standardized, there is a great benefit to the health care process that surpasses institutional and regional boundaries. Based on this, it is likely that the acceptability of the records may be a forgone conclusion. However, a minimum standard must be met to ensure information transportability when a different provider sees the patient.

The use of and acceptability of portable health records must also be based on trust. Patients in the Kenyan setting tend to take the physicians instructions as the only viable information. Hence, there are a great percentage of patients who may opt to use these records only if their personal clinician endorses them (Price et al. 2013). On the one hand, the physician must then trust the patient not alter or insert any other information other than that contained in the records. On the other hand, the patient must trust the physician to include all the relevant information (White and Danis 2013). In some situations, the patient may correct erroneous entries in their record, though. Additionally, from the patients' perspective, the records must be personalized with explanations that are clear for them, preferably in plain language. In addition, the record can have links to educational resources, decision aids, tools to promote action and periodic reminders (Price et al. 2013).

## 8.6   Portable Health Records and Accommodation/Adequacy

Adequacy/accommodation is one domain of access to health care. It refers to the organizational aspects of the health care. Access to care is limited if the distribution of resources is uneven across different levels of care. In the Kenyan setting, different services are received in different levels of care. This may compare to the specialized systems in the developed world like trauma centers. A well-organized health system means that it can easily accommodate the needs of their clients (Levesque et al. 2013). Hours of operation, proper referral systems, and facility structures denote a well-organized system.

In these settings, the use of portable health records can be promoted a little easier. Unfortunately, the distribution of resources in the Kenyan setting is not equitable. For example, the referral hospitals where all types of specialized services can be offered are located in urban areas and not accessible to all. Further, the referral systems are not well organized meaning that some patients will walk to the referral hospitals for care sans any medical records, because they refer themselves. In this scenario the use of portable health records is not supported. The characteristics of the health resources can impede or promote the portability of health records (Donabedian 1973).

## 8.7   Summary

Potable health records have the potential to improve availability, accessibility, affordability, acceptability, and adequacy of care. In addition, when well implemented, they can be affordable to the user and decrease the cost of care. However, issues of portability and control must be addressed. In the next section, we focus on how we can allow access to patients' records without jeopardizing their safety.

## References

Cushman R, Froomkin AM, Cava A, Abril P, Goodman KW. Ethical legal and social issues for personal health records and applications. J Biomed Inform. 2010;43:51–5.
Delbanco TJ, Walker SK, Bell JD, Darer JG, Elmore N, Farag HJ, Feldman R, Mejilla L, Ngo RJD. Inviting patients to read their doctors' notes: a quasi-experimental study and a look ahead. Ann Intern Med. 2012;157(7):461–70.

Detmer D, Bloomrosen M, Tang P. Integrated personal health records: transformative tools for consumer-centric care. BMC Med Inform Decis Mak. 2008;8:45. https://doi.org/10.1186/1472-6947-8-45.

Donabedian A. Aspects of medical care administration. 1st ed. Cambridge: Harvard University Press; 1973.

George T, Hopla DL. Advantages of personal health records. Nurs Crit Care. 2015;10(6):10–2. https://doi.org/10.1097/01.CCN.0000472852.70431.1b.

Getzen TE. Health economics and financing. 5th ed. Somerset: Wiley; 2013.

Han JE, Rabinovich M, Abraham P, Satyanarayana P, Liao TV, Udoji TN, et al. Effect of electronic health record implementation in critical care on survival and medication errors. Am J Med Sci. 2016;351(6):576–81.

Hlamka JD, Mandl KD, Tang PC. Early experiences with personal health records. J Am Med Inform Assoc. 2008;15(1):1–7. https://doi.org/10.1197/jamia.M2562.

Kahn JS, Aulakh V, Bosworth A. What it takes: characteristics of the ideal personal health record. Health Aff (Millwood). 2009;28(2):369–76. https://doi.org/10.1377/hlthaff.28.2.369.

Kerns JW, Krist AH, Longo DR, Kuzel AJ, Woolf SH. How patients want to engage with their personal health record: a qualitative study. BMJ Open. 2013;3(7):e002931. https://doi.org/10.1136/bmjopen-2013-002931.

Kraan CW, Piggott JJ, van der Vegt F, Wisse L. Personal health records: solving barriers to enhance adoption. E-health Strateg. 2015.

Lester M, Boateng S, Studeny J, Coustasse A. Personal health records: beneficial or burdensome for patients and healthcare providers? Perspect Health Inf Manag. 2016;13(Spring):1h.

Levesque FJ, Harris MF, Russel G. Patient-centered access to health care: conceptualizing access at the interface of health systems and populations. Int J Equity Health. 2013;12:18. https://doi.org/10.1186/1475-9276-12-18.

Merriam-Webster. Availability. n.d. https://www.merriam-webster.com/dictionary/availability. Accessed 21 Jul 2018.

Palen TE, Ross C, Powers JD, Xu S. Association of online patient access to clinicians and medical records with use of clinical services. JAMA. 2012;308(19):2012–9.

Peacock SA, Reddy SG, Leveille J, Walker TH, Payne NV, Elmore JG. Patient portals and personal health information online: perception, access, and use by US adults. J Am Med Inform Assoc. 2017;24(e1):e173–7.

Price MM, Pak R, Muller H, Stronge A. Older adults' perceptions of usefulness of personal health records. Univ Access Inf Soc. 2013;12:191–204.

Radhakrishna K, Goud BR, Kasthuri A, Waghmare A, Raj T. Electronic health records and information portability: a pilot study in a rural primary healthcare center in India. Perspect Health Inf Manag. 2014;11:1b.

Ralson J, Revere D, Robins L, Goldberg H. Patients' experience with a diabetes support programme based on an interactive electronic health record: qualitative study. Br Health J. 2004;328(7449):1159.

Ross S, Todd J, Moore L, Beaty B, Wittevrongel L, Lin CT. Expectations of patients and physicians regarding patient accessible health records. J Health Internet Res. 2005;7(2):e13.

Spil T, Klein R. Personal health records success: why Google Health failed and what does that mean for Microsoft HealthVault? In: Conference: proceedings of the 2014 47th Hawaii international conference on system sciences. IEEE; 2014. https://doi.org/10.1109/HICSS.2014.353.

Vodicka E, Mejilla R, Leveille SG, Ralston JD, Darer JD, Delbanco T, et al. Online access to doctors' notes: patient concerns about privacy. J Med Internet Res. 2013;15(9):e208.

Weitzman ER, Kaci L, Mandl KD. Acceptability of a personally controlled health record in a community health record in a community–based setting: implications for policy and design. J Med Internet Res. 2009;11(2):e14. https://doi.org/10.2196/jmir.1187.

White A, Danis M. Enhancing patient-centered communication and collaboration by using the electronic health record in the examination room. JAMA. 2013;309(22):2327–8.

Yau G, Williams A, Brown J. Family physicians' perspectives on personal health records: qualitative study. Can Fam Physician. 2011;57(5):178–84.

# Chapter 9
# Privacy

**Lovette Chinwah-Adegbola**

**Abstract**  The advent of the Health Insurance and Portability Accountability Act of 1996 (HIPAA) signaled a paradigm shift in how health records are stored, used, and released. The Act not only outlined provider responsibilities but also codified patients' rights. Now, in the more than two decades of its existence, providers as well as consumers have encountered challenges in their interpretation and implementation of HIPAA, especially the Privacy Rule component of the Act. This chapter will examine these challenges, and, hopefully, offer alternative approaches to conceptualizing privacy and its attendant variables. The chapter will conclude with a discussion of how communication pathways amongst the various entities—patients, doctors, hospitals, and agencies—can mitigate privacy concerns and complaints.

**Keywords**  Privacy rule · Privacy notice · Confidentiality · EHRs · Electronic health records · Privacy communication · Communication and privacy · HHS HIPAA

## 9.1  Background and Introduction

Almost gone are the days when patients lugged around hard copies of their medical records from one doctor's office to another doctor's office or had to wait for days or weeks to secure access to their medical records. Technological advances have made these tasks seamless among entities, and the bulky paper trails have been replaced, mostly by portable electronic health records (EHRs), thus allowing patients to access their medical records via multiple channels, and with the freedom to share such records as they please.

L. Chinwah-Adegbola (✉)
Department of Humanities, Central State University, Wilberforce, OH, USA
e-mail: Lchinwah@centralstate.edu

© Springer Nature Switzerland AG 2019
E. R. Onyejekwe et al. (eds.), *Portable Health Records in a Mobile Society*,
Health Informatics, https://doi.org/10.1007/978-3-030-19937-1_9

Naturally, privacy concerns have emerged. How safe are the records? How do authorized users get access to their records? Are timed access and automatic expiration enough? If enough, how robust are the parameters and protocols? How user friendly are the electronic health portals? Are authorized entities and consumers up-to-date in their interpretation of the Privacy Act? How would authorized users make sure that the data are available and/or released only to them? And, if not, with what consequences?

In an attempt to address some of these concerns, the United States Congress passed the Health Insurance and Portability Accountability Act of 1996 (HIPAA). HIPAA is within the purview of the U.S. Department of Health and Human Services (HHS), and a core component of HIPAA is the Privacy Rule, which is enforced by the Office of Civil Rights (OCR), a unit of HHS.

> The Rule requires appropriate safeguards to protect the privacy of personal health information and sets limits and conditions on the uses and disclosures that may be made of such information without patient authorization. The Rule also gives patients' rights over their health information, including rights to examine and obtain a copy of their health records, and to request corrections (U.S. Department of Health and Human Services 2018a).

The Rule serves as the foundation for the Notice of Privacy Practices that is given to patients—in person or by electronic or regular mail. Now, some providers are experimenting with a hybrid Notice—asking patients to read a mounted hard copy of the Notice and asking them to acknowledge the Notice by signing on an electronic pad, thereby bypassing the necessity for giving patients a hard copy for their records. This hybrid process typically unfolds this way: A patient arrives at the check-in desk. A staff member processes the patient's insurance and/or other relevant information and instructs the patient to acknowledge the acceptance of the Privacy Notice. The electronic version of the Notice is displayed on the staff's computer screen, and a mounted hard copy of the Notice is on the counter. When the patient is asked to electronically sign the signature-only pad, the assumption is that what's on the staff's computer screen is the same as the mounted copy. After the patient acknowledges receipt of the Privacy Notice by signing the signature-only pad, the staff asks the patient whether he or she wants a printed copy of the Notice, and it is up to the patient to accept or decline the offer. In other hybrid setups, the patient does not have access to the mounted hard copy and is asked only to sign the signature pad. Such inconsistencies and privacy concerns remain current. Sweeny analyzed the public records of hospitalization data in the state of Washington and how they were shared with other entities (Sweeny 2013). She discovered that a newspaper staff easily matched patient medical data with publicly available records when the staff member compared the public data set from the State's archives of hospitalizations with instances of identifiable patient names in news reports. This observation led Sweeny to ask, "Are privacy safeguards sufficient to protect patients from harm?"

First, I offer some definitions as a framework for understanding the subsequent discussion. Merriam Webster's Dictionary defines privacy as "freedom from unauthorized intrusion" (Merriam-Webster n.d.). Researchers at the University of California, Irvine, offer an expanded definition of privacy as "an individual's desire to control who has access to him/herself (sic) and whether there are "adequate provisions to protect the privacy interests of participants" (UCI Researchers 2011). According to the same researchers, confidentiality, however, is "how the participant's identifiable private information will be handled, managed, and disseminated" and whether there are "adequate provisions to maintain the confidentiality of data." In effect, "Privacy applies to person, and confidentiality applies to data."

The HHS defines confidentiality as "protecting information from unauthorized disclosure to people or processes" and privacy as "a set of fair information practices to ensure that an individual's personal information is accurate, secure, and current, and that individuals are informed how their personal data will be used" (U.S. Department of Health and Human Services 2018b).

To underscore the previous definitions, "Justice Samuel Dennis Warren and Justice Louis Brandeis define privacy as the right 'to be let alone'" (Ozair et al. 2015). Richard Rognehaugh, echoes the definitions offered by the aforementioned justices when he defines privacy "as the right of an individual to keep information about themselves from being disclosed to others; the claim of individuals to be let alone, from surveillance or interference from other individuals, organizations or the government" (Ozair et al. 2015).

The conflation of confidentiality with privacy sometimes leads to a semantic confusion on the part of providers and as well as consumers. Ozair et al. concludes that, "Although controlling access to health information is important, but is (sic) not sufficient for protecting the confidentiality" (Ozair et al. 2015).

The courts have maintained in numerous decisions that people should have a reasonable expectation of privacy in their lives. Along the same lines, some legal scholars argue that the Fourth and 14th Amendments in the Bill of Rights guarantee this expectation.

The Fourth Amendment, as it reads, guarantees

> The right of the people to be secure in their persons, houses, papers, and effects, against unreasonable searches and seizures, shall not be violated, and no warrants shall issue, but upon probable cause, supported by oath or affirmation, and particularly describing the place to be searched, and the persons or things to be seized (National Constitution Center 2018).

The 14th Amendment, as it reads, guarantees that

> All persons born or naturalized in the United States, and subject to the jurisdiction thereof, are citizens of the United States and of the State wherein they reside. No State shall make or enforce any law which shall abridge the privileges or immunities of citizens of the United States; nor shall any State deprive any person of life, liberty, or property, without due process of law; nor deny to any person within its jurisdiction the equal protection of the laws (National Constitution Center 2018).

The key phrase in the 14th Amendment, legal scholars argue, is the "due process" clause. A few HIPAA cases have been litigated on the bases of the Fourth and 14th

Amendments, with plaintiffs and respondents arguing that these Amendments guarantee people's rights to be left alone and not to be subjected to the disclosure of their private data. In Urbina v. Carson, 2007, the U.S. District Court, Eastern District of California, determined that the plaintiff could not prove that the defendant violated his Fourth Amendment rights when he disclosed his personal medical record, which, by the way, the plaintiff had provided to the defendant in the course of the discovery process in a lawsuit. In this case, however, the Court said that the plaintiff did not have a reasonable expectation of privacy in such instance (Fourth Amendment 2007).

Rodriguez asserts that the "HIPAA Security Rule requires that health care providers set up physical, administrative, and technical safeguards to protect your electronic health information," including access controls (passwords, PINs), encryption (providing a "key") and audit trail (monitors changes to the information and who accesses it) (Rodriguez 2011). Are these enough?

According to the Student Press Law Center, a privacy breach or an invasion of privacy has four components (SPLC 2011):

1. False Light occurs when someone is "unflatteringly portrayed—in words or pictures—a person as something that he or she is not." To put it another way, it is when a juxtaposition of facts or images results in misleading information about a person.
2. Misappropriation of Name or Likeness "is the unauthorized use of a person's name, photograph, likeness, voice or endorsement to promote the sale of a commercial product or service."
3. Public Disclosure of Private and Embarrassing Facts.
4. Intrusion occurs when someone "gathers information about a person in a place where that person has a reasonable right to expect privacy."

One can imagine that a breach of one or more of these components can have dire consequences for patients as well healthcare providers, as evident later in this chapter.

## 9.2   HIPAA and the Privacy Rule

The following are the key parameters of the Privacy Rule (U.S. Department of Health and Human Services 2018b):

1. Providers must develop privacy rules and give their patients Notice of Privacy Practices.
2. Providers must have a Privacy Official.
3. Providers must provide physical, administrative, and technical safeguards.

4. Patients/consumers/individuals have the right to review their records and limit the type and quantity of information shared. In addition, they have the right to be given a copy of such records if requested.
5. Providers may only "use, request, and disclose" the "minimum necessary" information on protected health information.
6. Providers must seek written consent from patients before they release their information.
7. Releasing protected information is subject to "need to know basis."
8. Providers must train their employees on their (providers) privacy policies and rules.

Though the Privacy Act represented a major shift in our conceptualization of how health records are maintained and used, it has not consistently resulted in effective and efficient communication among providers, consumers, compliance agencies, and other entities.

According to the complaints made to the Office of Civil Rights on HIPAA about the Privacy Rule, from 2013 (compliance year for providers) to 2016, the top issues in the cases that required corrective action by HHS included impermissible uses and disclosures, access, administrative safeguards, notice to individuals, and technical safeguards (U.S. Department of Health and Human Services 2018c) (Table 9.1).

As seen in Fig. 9.1, of these top issues, health information privacy complaints generally showed an upward trend from 2013 to 2016 (U.S. Department of Health and Human Services 2018d).

Seventy-four percent of the total investigated resolutions resulted in corrective actions.

**Table 9.1** Top five issues in investigated cases closed with corrective action, by calendar year

| Year | Issue 1 | Issue 2 | Issue 3 | Issue 4 | Issue 5 |
|------|---------|---------|---------|---------|---------|
| 2016 | Access | Impermissible Uses and Disclosures | Safeguards | Administrative safeguards | Technical safeguards |
| 2015 | Impermissible uses and disclosures | Safeguards | Access | Administrative safeguards | Technical safeguards |
| 2014 | Impermissible uses and disclosures | Safeguards | Access | Administrative safeguards | Technical safeguards |
| 2013 | Impermissible uses and disclosures | Safeguards | Access | Minimum necessary | Administrative safeguards |

Note: Retrieved from The Department of Health and Human Services. Health information privacy, top five issues in investigated cases closed with corrective action, by calendar year. 2018

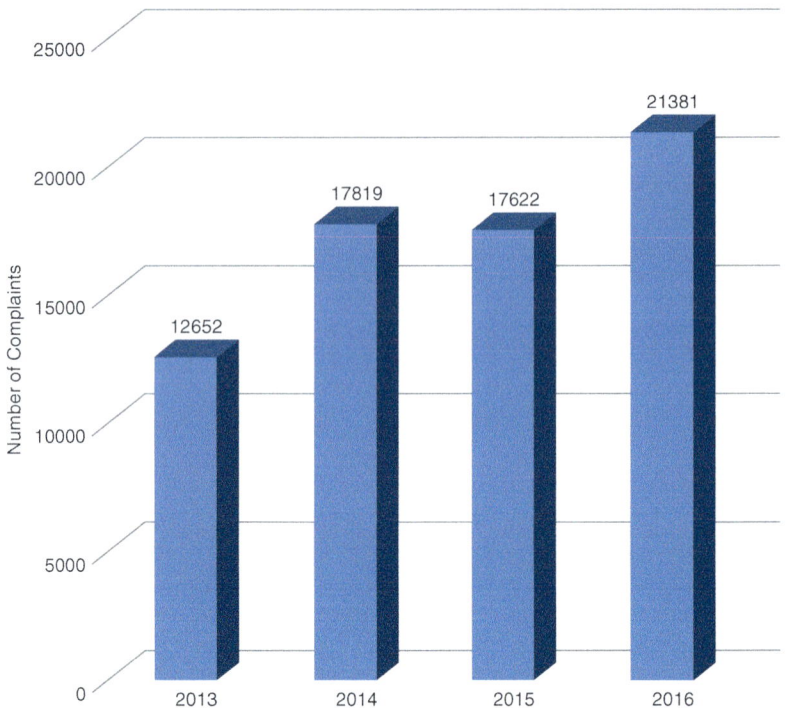

**Fig. 9.1** Health information privacy complaints received by calendar year (used HHS data to create chart). Note: Adapted from The Department of Health and Human Services. Health information privacy complaints received by calendar year. 2018

**Table 9.2** Outcome of complaint investigations, by calendar year

| Year | Investigated: no violation (number/ percentage) | | Resolved after intake and review (number/ percentage) | | Investigated: corrective action obtained (number/ percentage) | | Technical assistance (number/ percentage) | | Total resolutions |
|---|---|---|---|---|---|---|---|---|---|
| 2013 | 994 | 7% | 6917 | 49% | 3470 | 25% | 2753 | 19% | 14,134 |
| 2014 | 668 | 4% | 10,401 | 59% | 1288 | 7% | 5128 | 29% | 17,485 |
| 2015 | 360 | 2% | 12,634 | 72% | 730 | 4% | 3817 | 22% | 17,541 |
| 2016 | 204 | 1% | 16,780 | 70% | 706 | 3% | 6204 | 26% | 23,894 |

Note: Adapted from HHS. Health information privacy, enforcement results by year. 2018

Table 9.2 shows the outcome of complaint investigations. Figures 9.2 and 9.3 show total resolutions and total investigations respectively (U.S. Department of Health and Human Services 2018e).

## 9.3 HHS Case Examples

The US Department of Health and Human Services has shared hundreds of cases or examples that mirror the aforementioned top issues. Ten of those cases are quoted verbatim in this chapter. In each case, the HHS summarizes the incident, provides

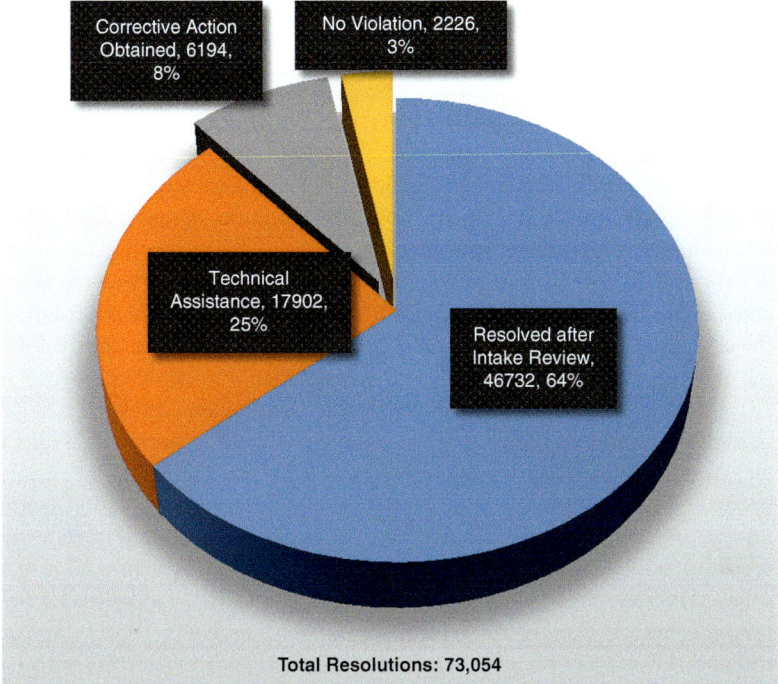

**Fig. 9.2** Total resolutions: 2013 through 2016 (used HHS data to create chart). Note: Adapted from HHS. Health information privacy, enforcement results by year. 2018

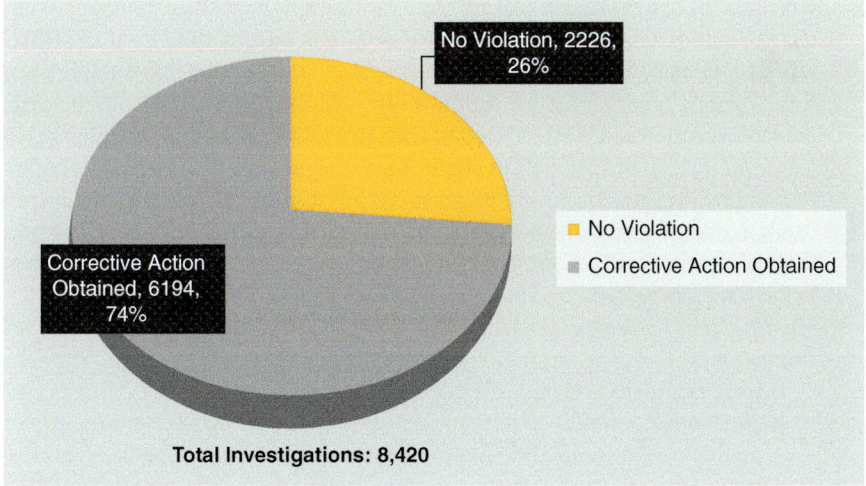

**Fig. 9.3** Total investigations enforcement results: 2013 through 2016 (used HHS data to create chart). Note: Adapted from HHS. Health information privacy, compliance enforcement results by year. 2018

or suggests a resolution, and identifies the corrective action that was taken to prevent or reduce the recurrence of the incident (U.S. Department of Health and Human Services n.d.).

1. *Hospital Implements New Minimum Necessary Polices for Telephone Messages*
   Covered Entity: General Hospital

   Issue: Minimum Necessary; Confidential Communications

   A hospital employee did not observe minimum necessary requirements when she left a telephone message with the daughter of a patient that detailed both her medical condition and treatment plan. An OCR investigation also indicated that the confidential communications requirements were not followed, as the employee left the message at the patient's home telephone number, despite the patient's instructions to contact her through her work number. To resolve the issues in this case, the hospital developed and implemented several new procedures. One addressed the issue of minimum necessary information in telephone message content. Employees were trained to provide only the minimum necessary information in messages and were given specific direction as to what information could be left in a message. Employees also were trained to review registration information for patient contact directives regarding leaving messages. The new procedures were incorporated into the standard staff privacy training, both as part of a refresher series and mandatory yearly compliance training.

2. *HMO Revises Process to Obtain Valid Authorizations*
   Covered Entity: Health Plans/HMOs

   Issue: Impermissible Uses and Disclosures; Authorizations

   A complaint alleged that an HMO impermissibly disclosed a member's PHI, when it sent her entire medical record to a disability insurance company without her authorization. An OCR investigation indicated that the form the HMO relied on to make the disclosure was not a valid authorization under the Privacy Rule. Among other corrective actions to resolve the specific issues in the case, the HMO created a new HIPAA-compliant authorization form and implemented a new policy that directs staff to obtain patient signatures on these forms before responding to any disclosure requests, even if patients bring in their own "authorization" form. The new authorization specifies what records and/or portions of the files will be disclosed and the respective authorization will be kept in the patient's record, together with the disclosed information.

3. *Mental Health Center Corrects Process for Providing Notice of Privacy Practices*
   Covered Entity: Outpatient Facility

   Issue: Notice

   A mental health center did not provide a notice of privacy practices (notice) to a father or his minor daughter, a patient at the center. In response to OCR's investigation, the mental health center acknowledged that it had not provided the complainant and his daughter with a notice prior to her mental health evaluation. To resolve this matter, the mental health center revised its intake assessment policy and procedures to specify that the notice will be provided, and the clinician will attempt to obtain a signed acknowledgement of receipt of the notice prior to the intake assessment. The acknowledgement form is now included in the intake package of forms. The center also provided OCR with written assurance that all policy changes were brought to the attention of the staff involved in the daughter's care and then disseminated to all staff affected by the policy change.

4. *Entity Rescinds Improper Charges for Medical Record Copies to Reflect Reasonable, Cost-Based Fees*

   Covered Entity: Private Practice

   Issue: Access

   A patient alleged that a covered entity failed to provide him access to his medical records. After OCR notified the entity of the allegation, the entity released the complainant's medical records but also billed him $100.00 for a "records review fee" as well as an administrative fee. The Privacy Rule permits the imposition of a reasonable cost-based fee that includes only the cost of copying and postage and preparing an explanation or summary if agreed to by the individual. To resolve this matter, the covered entity refunded the $100.00 "records review fee."

5. *Private Practice Implements Safeguards for Waiting Rooms*

   Covered Entity: Private Practice

   Issue: Safeguards; Impermissible Uses and Disclosures

   A staff member of a medical practice discussed HIV testing procedures with a patient in the waiting room, thereby disclosing PHI to several other individuals. Also, computer screens displaying patient information were easily visible to patients. Among other corrective actions to resolve the specific issues in the case, OCR required the provider to develop and implement policies and procedures regarding appropriate administrative and physical safeguards related to the communication of PHI. The practice trained all staff on the newly developed policies and procedures. In addition, OCR required the practice to reposition its computer monitors to prevent patients from viewing information on the screens, and the practice installed computer monitor privacy screens to prevent impermissible disclosures.

6. *Large Medicaid Plan Corrects Vulnerability that Resulted in Disclosure to Non-BA Vendors*

   Covered Entity: Health Plans

   Issue: Impermissible Uses and Disclosures; Safeguards

   A municipal social service agency disclosed protected health information while processing Medicaid applications by sending consolidated data to computer vendors that were not business associates. Among other corrective actions to resolve the specific issues in the case, OCR required that the social service agency develop procedures for properly disclosing protected health information only to its valid business associates and to train its staff on the new processes. The new procedures were instituted in Medicaid offices and independent health care programs under the jurisdiction of the municipal social service agency.

7. *Health Sciences Center Revises Process to Prevent Unauthorized Disclosures to Employers*

Covered Entity: General Hospitals

Issue: Impermissible Uses and Disclosures; Authorizations

A state health sciences center disclosed protected health information to a complainant's employer without authorization. Among other corrective actions to resolve the specific issues in the case, including mitigation of harm to the complainant, OCR required the Center to revise its procedures regarding patient authorization prior to release of protected health information to an employer. All staff was trained on the revised procedures.

8. *National Pharmacy Chain Extends Protections for PHI on Insurance Cards*

Covered Entity: Pharmacies

Issue: Impermissible Uses and Disclosures; Safeguards

A pharmacy employee placed a customer's insurance card in another customer's prescription bag. The pharmacy did not consider the customer's insurance card to be protected health information (PHI). OCR clarified that an individual's health insurance card meets the statutory definition of PHI and, as such, needs to be safeguarded. Among other corrective actions to resolve the specific issues in the case, the pharmacy revised its policies regarding PHI and retrained its staff. The revised policies are applicable to all individual stores in the pharmacy chain.

9. *Private Practice Revises Process to Provide Access to Records*

Covered Entity: Private Practices

Issue: Access

A private practice failed to honor an individual's request for a complete copy of her minor son's medical record. OCR's investigation determined that the private practice had relied on state regulations that permit a covered entity to provide a summary of the record. OCR provided technical assistance to the covered entity, explaining that the Privacy Rule permits a covered entity to provide a summary of patient records rather than the full record only if the requesting individual agrees in advance to such a summary or explanation. Among other corrective actions to resolve the specific issues in the case, OCR required the covered entity to revise its policy. In addition, the covered entity forwarded the complainant a complete copy of the medical record.

10. *Private Practice Revises Process to Provide Access to Records Regardless of Payment Source*

Covered Entity: Private Practices

Issue: Access

At the direction of an insurance company that had requested an independent medical exam of an individual, a private medical practice denied the individual a copy of the medical records. OCR determined that the private practice denied the individual access to records to which she was entitled by the Privacy Rule.

Among other corrective actions to resolve the specific issues in the case, OCR required that the private practice revise its policies and procedures regarding access requests to reflect the individual's right of access regardless of payment source.

## 9.4 Analysis and Conclusion

A review of these cases reveals a pattern. The complaints share a common denominator: failure to communicate efficiently and effectively. The root of the word "communicate" is *communicare, Latin for* "making something common." In essence, the sender and the receiver must arrive at a common or shared meaning of the message. Many of the complaints, including the ones described in this chapter, arose from a misinterpretation of the tenets of the Privacy Rule, by providers and customers alike. Clearly, technological issues play a vital role as well. Timed access, automatic expirations, PINs, and encryption certainly mitigate compliance issues, and, perhaps, assure, to some extent, compliance with the Rule. In addition, the understanding that a consumer or provider brings to the privacy compliance continuum is equally important.

For example, in some cases, there is a lack of shared understanding among parties as to what constitutes protected health information or authorized user. In other cases, the concepts of privacy and confidentiality are comingled, thereby causing further confusion. The data also show that challenges are present in both private and public settings, and that the complaints straddle the core components of the Privacy Rule.

The OCR is responding to complaints on the Rule by providing online training and other resources to healthcare providers. It is imperative, however, that there be a sharing of meaning among all entities, and this can be effected through a robust communication feedback loop.

Effective communication pathways that include a continual feedback loop could possibly lessen the misinterpretation and application of the Privacy Rule. Patients should be empowered to fully understand the components and meaning of the terms in the privacy notices that they receive, the circumstances under which their health records may be disclosed, and the consequences of such disclosure. In this instance, providers are encouraged to strengthen training workshops not just for their employees but also for patients (possibly done during the initial intake process).

The proposed communication pathway is grounded in Schramm's (Sage Knowledge 2018) components of the communication process: sender, receiver, message, channel, context, noise, and feedback. Providers and patients need to understand the characteristics of these components and how they are interrelated.

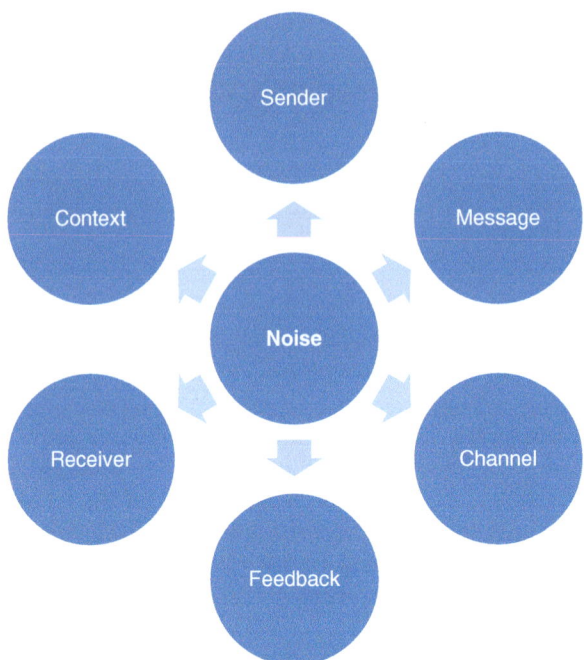

Sender: The sender conceptualizes what he or she wishes to convey using his or her experiences. Does the sender know whether an acknowledgement form is necessary?

Message: The sender creates a message based on his or her own language set, experiences, and interpretation of the intended meaning of his or her message. For example, are there jargons or terms that might be foreign to patients? Yes, many healthcare providers have interpreters, but would such interpretation be lost in translation? Who is an authorized user? What is protected health information? What is a reasonable cost-based fee for transcribing and/or compiling health records? What constitutes valid authorization?

Channel: The sender selects the medium through which he or she sends the message. The channel must be appropriate to the audience and the context. For example, is it enough that only the hard copy versions of the Privacy Notice are made available to patients? Could patients benefit from an audio-visual explanation of the Notice? Would the posting of the audio-visual version in the patients' EHR portal be useful to patients?

Receiver: The receiver has to make sense of the message using his or her own language set and experiences. To what extent do patients and their providers share the same meaning of the parameters of the Privacy Notice and Rule?

Context: The context refers to the environment in which the communication occurs, which can impact the receptivity of the sender and the receiver. Sometimes, patients are waiting in line with other patients as they are informed of the Privacy Notice. Do healthcare providers understand the invasion of privacy issues inherent in exposing computer screen to bystanders? It would be prudent, in this context, for healthcare providers to equip their computers with privacy screens. Also, could there be a different approach to determining the setting in which patients receive the Notice? Patients receive hard copies of the Privacy Notice via regular mail unless they opt out, but how many who receive the Notices read them? How much health information should be shared?

Noise: Noise is anything that interferes with the communication pathway. Noise can also include physiological, technological, and other variables and can occur in any of the other six components. "Noise" in the case examples applies to semantic variables—meanings and interpretations of the terms within the Privacy Notice.

Feedback: The key component in this communication pathway is the *feedback loop*. It is the feedback loop that reduces or eliminates the amount and frequency of the noise inherent in communication between patients and providers. In other words, there must be a continual dialogue and clarification of terms among all parties—patients, providers, and agencies. The protocols for applying the Privacy Rule and Notice must undergo periodic review, and healthcare providers would do well to seek feedback from patients on those protocols.

Future work could explore the perspectives of the providers to determine the communication barriers that they feel often stand in the way of the accurate interpretation of the Rule and Notice by patients. Such cross-referencing would be valuable to our understanding of how the Privacy Rule and its attendant Privacy Notice are interpreted and implemented by all entities.

# References

Fourth Amendment. HIPAA privacy violation is not a fourth amendment issue. 2007. http://fourth-amendment.com/?p=1433. Accessed 22 Jul 2018.

Merriam-Webster. Privacy. n.d.. https://www.merriam-webster.com/dictionary/privacy. Accessed 23 Jan 2018.

National Constitution Center. Amendment. 2018. https://constitutioncenter.org/interactive-constitution/amendments/. Accessed 22 Jul 2018.

Ozair F, Jamshed N, Sharma A, Aggarwal P. Ethical issues in electronic health records: a general overview. In: Perspectives in clinical research. 2015. https://www.ncbi.nlm.nih.gov/pmc/articles/PMC4394583. Accessed 22 Jul 2018.

Rodriguez L. Privacy, security, and electronic health records. 2011. https://www.healthit.gov/buzz-blog/privacy-and-security-of-ehrs/privacy-security-electronic-health-records/

Sage Knowledge. The Schramm model of communication. 2018. http://sk.sagepub.com/books/key-concepts-in-marketing/n46.xml. Accessed 25 July 2018.

SPLC. Invasion of privacy laws, in brief. 2011. http://www.splc.org/article/2011/06/invasion-of-privacy-law. Accessed 22 Jul 2018.

Sweeny L. Matching known patients to health records in Washington State data. 2013. https://dataprivacylab.org/projects/wa/1089-1.pdf. Accessed 22 Jul 2018.

U.S. Department of Health and Human Services. The HIPAA privacy rule. 2018a. https://www.hhs.gov/hipaa/for-professionals/privacy/index.html. Accessed 22 Jul 2018.

U.S. Department of Health and Human Services. Summary of the HIPAA privacy rule. 2018b. https://www.hhs.gov/hipaa/for-professionals/privacy/laws-regulations/index.html. Accessed 22 Jul 2018.

U.S. Department of Health and Human Services. Top five issues in investigated cases closed with corrective action, by calendar year. 2018c. https://www.hhs.gov/hipaa/for-professionals/compliance-enforcement/data/top-five-issues-investigated-cases-closed-corrective-action-calendar-year/index.html. Accessed 22 Jul 2018.

U.S. Department of Health and Human Services. Health information privacy complaints, received by calendar year. 2018d. https://www.hhs.gov/hipaa/for-professionals/compliance-enforcement/data/complaints-received-by-calendar-year/index.html. Accessed 22 Jul 2018.

U.S. Department of Health and Human Services. Enforcement results by year. 2018e. https://www.hhs.gov/hipaa/for-professionals/compliance-enforcement/data/enforcement-results-by-year/index.html. Accessed 22 Jul 2018.

U.S. Department of Health and Human Services. HHS case examples. n.d. https://www.hhs.gov/hipaa/for-professionals/compliance-enforcement/examples/all-cases/index.html. Accessed 22 Jul 2018.

UCI Researchers. Privacy vs. confidentiality: what is the difference? 2011. https://research.uci.edu/compliance/human-research-protections/docs/privacy-confidentiality-hrp.pdf. Accessed 22 Jul 2018.

# Part IV
# Delivery

# Chapter 10
# Patient-Provider Communication

**LaQuasha Gaddis**

**Abstract** Communication between providers and patients is a very aspect important of health services and patient care. When such communication is patient-centered, it can contribute to fostering healing relationships, enabling patient self-management, and making shared health decisions. Patient-provider communication is affected by several factors such as the patients' health and background, clinicians' personality and orientation, the health system or even the health system factors. Understanding the importance of communication and potential communication barriers helps in identifying new methods for improving communication and potentially improving the overall health outcomes.

**Keywords** Patient-provider communication · Provider communication · Patient communication · Health literacy · Patient engagement · Shared-decision making Digital technology

## 10.1 Patient-Provider Communication (PPC)

In a patient care setting, communication from patient to healthcare provider and from healthcare provider to patient (denoted as patient-provider in the sequel) is essential for high quality healthcare outcomes. Clear communication is the foundation for patients to be able to understand and act on health information provided by the provider, as well as to establish trust among their provider (Indian Health Service 2016). Patient communication similarly allows the healthcare provider to better understand the patient's health condition or needs and to develop a relationship necessary to provide effective health care solutions including self-management of chronic illnesses. Ha et al. states that a doctor's communication and interpersonal skills should encompass the ability to gather information to help

L. Gaddis (✉)
Centers for Disease Control and Prevention, Atlanta, GA, USA
e-mail: laquasha.gaddis@waldenu.edu

© Springer Nature Switzerland AG 2019
E. R. Onyejekwe et al. (eds.), *Portable Health Records in a Mobile Society*,
Health Informatics, https://doi.org/10.1007/978-3-030-19937-1_10

facilitate accurate diagnosis, counsel appropriately, and give therapeutic instructions (Ha et al. 2010). Patients who communicate effectively with their healthcare provider may be able to express how they feel about their health and medical treatment, as well as follow advice on making healthier lifestyle choices. This chapter will briefly discuss the different aspects that can have an impact on patient-provider communication including communication style, health literacy, race, and technology. The chapter will also discuss how patient-provider communication can be measured and various challenges that arise to reduce its effectiveness.

## 10.2 Patient-Provider Communication

Along with establishing a caring relationship with patients, the ultimate goal for a healthcare provider is to achieve the best outcome and patient satisfaction for the effective delivery of health care. Ideas about health and health behaviors are shaped through effective communication. A patient's perceptions of the quality of the healthcare received are highly dependent on the quality of their interactions with their health provider (Zill et al. 2014). The connection between patients and health providers can improve adherence to treatment, patient self-management, and shared decision-making. Although people have the ability to manage their health and health behaviors, the increasing complexity of health information requires the need for additional information, skills and supportive relationships that will help them to meet their health needs (HealthyPeople.gov 2016). Effective patient-provider communication can potentially help to regulate patients' emotions and facilitate understanding of medical information. Doctors can further identify patients' needs, perceptions, and expectations of their health issues (Ha et al. 2010).

Moreover, patient-centered encounters result in better patient and doctor satisfaction. The Institute of Healthcare Communication stated that a doctor's ability to explain, listen, and emphasize can have a positive impact on health outcomes, as well as, on patient satisfaction and experience of care (Institute for Healthcare Communication 2016). Creating a good interpersonal relationship, facilitating the exchange of information, and including patients in decision-making are considered the three main goals of patient-provider communication (Ha et al. 2010). An effective patient-provider relationship can further be considered a source of incentive, reassurance, motivation, and support. Consequently, satisfied patients are more likely to continue the patient-provider relationship and maintain recommended changes in health behaviors. The patient-provider relationship can help reinforce self-efficacy, job satisfaction, and bring about a positive view of a patient's health status (Ha et al. 2010). Greater patient satisfaction also leads to a reduction in healthcare costs due to improved health outcomes.

## 10.3   Measuring Patient-Centered Communication

Patient-centered communication (PCC) can be measured by direct observations, interviews, or surveys. The most frequently used measures are surveys of patients about their experience of care (Levinson et al. 2010). The Consumer Assessment of Healthcare Providers and Systems (CAHPS) and the Picker Foundation Survey are examples of surveys used to assess patient-centered care. Although PCC encompasses a variety of behaviors and attitudes, one challenge in measuring such a construct is how to gather information about communication behaviors and understanding their effects from both objective and subjective viewpoints. Another measuring challenge is the lack of theoretical and conceptual clarity (Epstein et al. 2015). Therefore, it is imperative to develop clear theory-based operational definitions and make sure there is clarity in how PCC is being measured. Measures should account for the communication behaviors of both the patient and practitioner. However, Epstein et al. stated that caution should be used when interpreting patient ratings of their doctors since bias can be created caused by other related constructs such as trust, self-efficacy, and satisfaction (Epstein et al. 2015).

## 10.4   Communication Style

During medical visits, patients and providers may process given health information differently based on how it is communicated. A patient's health outcome is dependent on how successful the processing of communication is, and a doctor who encourages open communication may obtain more complete health information about a patient's medical condition than a doctor who communicates one-way to the patient (American College of Obstetricians and Gynecologists 2014). The communication style of a health provider therefore influences a patient's attitude and perception regarding their health status, as well as, patient satisfaction with healthcare. Four aspects of communication style and satisfaction are: determining if healthcare providers listen carefully, explain things in such a way that patients will understand, show respect for what patients have to say, and spend enough time interacting with their patients (Jensen et al. 2011). Patient-centeredness is considered one type of communication style that is useful for health providers when interacting with patients to help produce better health outcomes (Fig. 10.1).

Patient-centeredness focuses on the patient and sees them as a person with a unique personal history and individual needs (Verlinde et al. 2012). It also provides care that is respectful of and responsive to a patient's preferences and values and ensures the patient values guide all clinical decisions. Moreover, the communication style of patient-centeredness implies high levels of caring and sharing. Research has demonstrated that a caring communication style aims at developing and maintaining a good relationship between the patient and provider, and also conveys friendliness, empathy, interest, and a desire to help (Cousin et al. 2012). Caring can

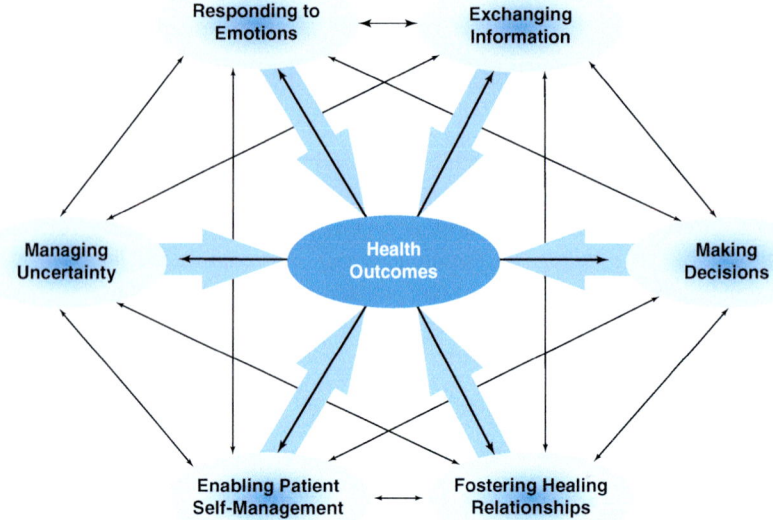

**Fig. 10.1** Patient-centered communication. Reprinted from "Patient-centered communication in cancer care: promoting healing and reducing suffering," Epstein RM, Street RL Jr. Patient-Centered Communication in Cancer Care: Promoting Healing and Reducing Suffering. National Cancer Institute, NIH Publication No. 07-6225. Bethesda, MD, 2007, p. 18

be further shown by the provider through verbal behaviors such as using statements of reassurance and support, positive reinforcement, and actively listening. By adapting a communication style to the patient's needs, the health provider can build a partnership and trust to help empower the patient to become more active in decision-making (Verlinde et al. 2012; Cousin et al. 2012).

## 10.5   Shared Decision-Making

Another important aspect of patient-provider communication is involving patients in their healthcare decisions, also known as shared decision-making. Shared-decision making is defined as a process where both the patient and provider share information, express treatment preferences, and agree on a treatment plan (American College of Obstetricians and Gynecologists 2014). By using shared-decision making, health providers can offer options and describe risks and benefits of each option while the patients can express their preferences and values (Barry and Edgman-Levitan 2012). Providers can further help their patients become more compliant with treatment and improve self-management of their medical conditions. Moreover, through patient-centered care, health providers can use this as a way of measuring patient perceptions, as well as, to assess the effectiveness of the care they are receiving (Rickert 2012).

## 10.6    The Influences on Patient-Provider Communication

Patient-provider communication is multidimensional and influenced by many factors including the severity of illness, culture, the health system, socioeconomic status, health literacy, and clinician factors as shown in Fig. 10.2 (Epstein et al. 2015). If these factors are not handled correctly or taken into consideration by a health provider, discordance in communication can be caused with the patient. As an example, the patient-provider relationship might suffer from social inequalities. Verlinde et al. conducted a literature review on the effects of social class and patient-provider communication (Verlinde et al. 2012). Social class was determined by the patient's income, education, or occupation. The review concluded that patients from a higher social class generally had better communication with their health provider and received more information, as well as, actively participated in shared decision-making than those from a lower social class (Barry and Edgman-Levitan 2012). Patients within a lower social class asked fewer questions, received less health information, and were more often dissatisfied with the communication style of their health provider. This provides further evidence that doctors may tend to communicate differently with patients based on their

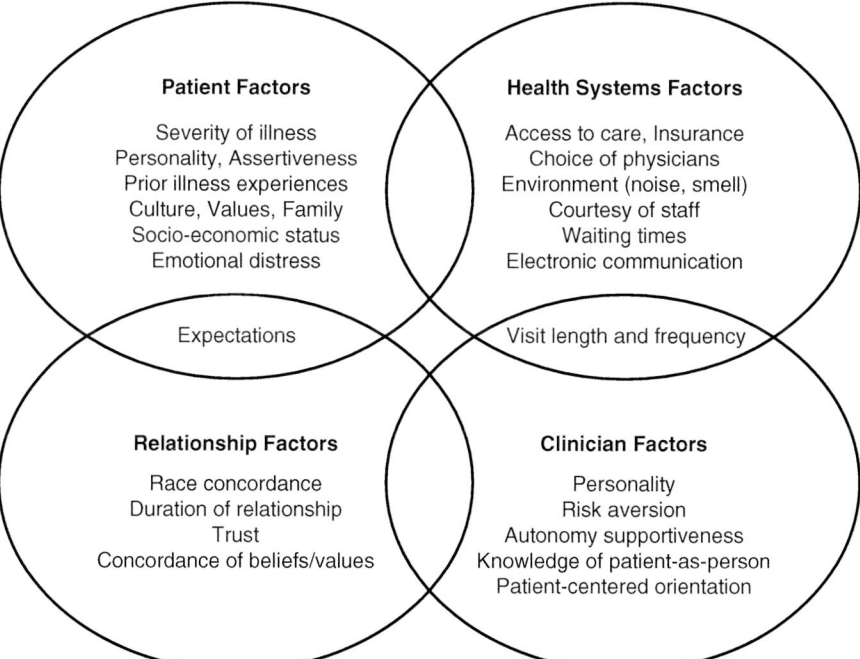

**Fig. 10.2**  Factors influencing patient-centered communication. Adapted from "Measuring patient-centered communication in Patient-Physician consultations: Theoretical and practical issues," by Epstein et al., 2005, Social Science & Medicine 61, p. 1517

socioeconomic status, and patients may adapt a different communication style according to their social class (Barry and Edgman-Levitan 2012).

Likewise, the choice of healthcare providers such as physician gender plays an essential role in patient-provider communication. While knowledge, clinical competence and experience in practice are important, some individuals look for the provider that will show greater empathy and support during their medical visit. Research has also shown that differences in gender are related to differences in communication style. For instance, female doctors tend to demonstrate a more partnership building behavior and provide more verbal encouragement than male doctors (Janssen and Lagro-Janssen 2012; Jefferson et al. 2013). Studies have further shown that female doctors also conduct longer consultations and are more empathetic (Jefferson et al. 2013). As previously mentioned, it is important for doctors and physicians to actively listen, as well as, be empathetic to their patients and provide the necessary support and health information that will improve health outcomes and health behaviors.

Jansen et al. conducted a literature review on women seeking gynecological or obstetrical care and the association of physician's gender to differences in communication style (Janssen and Lagro-Janssen 2012). Gender was an important factor in more that 75% of patients who preferred a female gynecologists-obstetrician. Female physicians also appeared to use a more patient-centered communication style compared to male physicians, which in turn has been shown to improve patient satisfaction (Janssen and Lagro-Janssen 2012). Moreover, female physicians have been found to be significantly more important for ethnic minority patients including Asian, Black, and mixed races compared to white patients (Jefferson et al. 2013). Nevertheless, the communication that takes place between a provider and a patient whether male or female is integral to the care patients receive.

## 10.7   The Effects of Trust, Race, and Ethnicity

Trust and communication are important factors for individuals when dealing with health care management of chronic illnesses. Patient-provider communication allows for trust and understanding to be built between the patients and providers. Additionally, the trust built through patient-provider communication can be associated with better satisfaction with healthcare services, an increase in treatment adherence, as well as, improved health outcomes. Abel et al. stated that a patient's trust in their health provider implies confidence that the provider's words are honest, and actions are appropriate in the provision of care and treatment (Abel and Efird 2013). However, lower levels of trust in health providers are also associated with lower patient satisfaction among minority populations including African Americans (Martin et al. 2013). Research has shown that patients lose trust in providers who do not listen to their concerns, care about their health issues, or provide enough information regarding treatment (Levinson et al. 2010).

It may be assumed by an individual that during a medical visit, their health provider is giving accurate information and has the patient's best interest in mind.

Providers may also believe that they know everything about their patient and health care needs; yet, lack the communication skills that will help improve health outcomes and patient satisfaction (Rickert 2012). A lack of trust and ineffective communication can lead to medical errors, misdiagnosis, or patient harm (Dingley et al. 2008). Some health providers lack the adequate training needed to provide quality care to patients. Thus, to be trustworthy, providers need to show better caring behaviors, good interpersonal communication skills, and a desire to promote the health of the patients they serve (Abel and Efird 2013).

Moreover, research has shown that provider communication behaviors can have varying effects on patient trust, depending on the patient's race. For example, minority groups including African Americans, are less likely to seek or receive needed services, such as procedures or routine treatments for common health conditions or chronic illnesses in comparison to Caucasians (Martin et al. 2013). Another study found that hypertensive Black women who trusted their health providers were more likely to be adherent to treatment versus those who did not trust their providers (Abel and Efird 2013) The lack of trust and communication can further increase health disparities among races. The distrustful relationships between African Americans and health care providers have also been influenced by past historical events such as the Tuskegee experiment. For instance, in the past African Americans were frequently used in experiments by White doctors to help improve medical and surgical techniques before attempting the same procedures on Whites (Abel and Efird 2013).

## 10.8  Healthy Literary and Patient-Provider Communication

Health literacy allows individuals to comprehend, evaluate and use health information to make informed choices that will increase their quality of life. However, an estimated 90 million Americans have problems understanding health information and struggle to maintain their health (American Health Information Management Association 2016; Tamura-Lis 2013). Health literacy can affect how people adhere to their medication. Individuals with low health literacy tend not to adhere to treatments, have increased medical care cost, and do not comprehend their medical conditions (Bailey et al. 2013). Likewise, physicians who do not recognize the low health literacy of individuals can contribute to patients' poor adherence by prescribing complex treatments and failing to explain the benefits or side effects of medications (Schoenthaler et al. 2009).

Health literacy is defined by The Centers for Disease Control and Prevention as the degree to which an individual has the capacity to obtain, communicate, process, and understand basic health information and services to make appropriate health decisions (Center for Disease Control and Prevention 2015). Health literacy can be considered a predictor of health and goes beyond just annual visits to the provider or routine screenings. Health literacy is about empowering people to better understand prescriptions, discharge instructions, consent forms, requests for health information, as well as, the ability to understand complex health services (American Health Information Management Association 2016). It cannot be assumed that because an individual can easily read a prescription that they also understand its effects and will take the medication as prescribed. Colbert et al. stated in their research study that health literacy has emerged as a potential variable linked to chronic disease management (Colbert et al. 2013).

## 10.9    The Impact of Low Health Literacy

Health literacy can affect how an individual communicates with their health provider or adherence to the recommended treatment if diagnosed with a chronic illness. A health provider may not take into consideration the individual's lifestyle, the cost of medication as well as the lack of knowledge or education regarding their health issues due to the poor health literacy of the patient (Schoenthaler et al. 2009). Thus, patient-provider communication is essential for patient health, knowledge, decision-making, and motivation (Nouri and Rudd 2015). Minorities, elderly, and individuals with low socioeconomic status or education are usually associated with having low health literacy. Low health literacy can further result in medication errors, reduced use of preventative services, unnecessary emergency room visits, ineffective management of chronic conditions, and higher mortality rates (Center for Health Care Strategies 2013). Individuals are also be unlikely to complete follow-up appointments and fail to thoroughly complete medical forms. Consequently, a healthcare provider must take time to understand patients' knowledge and perceptions and be able to address their uncertainty. Any educational content or communication should be appropriately matched to the patient's level of health literacy and aligned with their readiness to make a behavioral change (Zullig et al. 2013).

## 10.10    Addressing Low Health Literacy

Solutions to addressing low health literacy include increasing health education training for both health professionals and patients. It is also suggested to write health pamphlets and brochures in plain language to make them more understandable for individuals. According to Health.gov, plain language is a strategy for making written and oral information easier to understand and a tool used to improve health literacy (Health.gov n.d.). Another strategy is to improve patient-provider communication by creating respectful environments and keeping the patient engaged in the conversation (Center for Health Care Strategies 2013). By utilizing the "teach back" method, the health provider can confirm that the patient has an understanding of the health information that has been given to them.

## 10.11   The Influence of Technology

As healthcare continues to rapidly change, patient-provider communication has become fueled by the information revolution. The use of technology has become a practical way to disseminate health information. New platforms for integrating medical care and digital technology have been developed to enhance patient-provider communication (Gupta and Sheikh 2014). Such technological tools include electronic medical records (EMR), email, smartphone apps, telemedicine, and patient portals. With an ever growing and rapidly changing healthcare delivery system, it is necessary for providers to adapt to these changes that will help provide an efficient level of care to patients (Gupta and Sheikh 2014). Likewise, the way health information is delivered can have a positive impact on healthcare services, as well as, improve patients' compliance with treatments (HealthyPeople. gov 2016).

Before the advancement of technology, traditional method used by providers to communicate and interact with their patients regarding health information was primarily through face-to-face or telephone communication. However, digital technology has emerged as another viable way to improve patient-provider communication (Ye et al. 2010). Access to the internet via smartphones, laptops, or tablets give patients the opportunity to obtain health information, research treatment options, disease prognosis, electronic health records, and ask questions directly to the provider (Gupta and Sheikh 2014).

According to Caligtan et al. an estimated 79% of U.S. adults over the age of 18 years use the Internet, which is also used to send and receive email (Caligtan et al. 2012). Email communication has the potential to improve patient-provider communication by keeping patients actively involved in self-management of their care. For providers, communicating by email has been seen as a more convenient and easier way to exchange information, conduct a follow-up to patients' questions, and document appointments (Gupta and Sheikh 2014; Ye et al. 2010). Another form of digital communication is through short message service (SMS) that is useful for sending appointment reminders, data results, and other health education messages. In a study conducted by Hamine et al. SMS was considered the most commonly used tool and primary platform for facilitating patient-provider communication (Hamine et al. 2015).

## 10.12   Challenges to Using Digital Technology

Although digital technology has been proven as a useful tool for sending and receiving health information, it can also create a barrier to patient-provider communication. For example, electronic devices can be stressful and overwhelming for individuals such as the elderly who have limited experience using the Internet or sending an email (HealthyPeople.gov 2016; Hamine et al. 2015). Most importantly, both patients and providers share the concern of maintaining the privacy and confidentiality of their communication especially through email (Ye et al. 2010).

## 10.13   Medication Adherence and Patient-Provider Communication

Jensen et al. stated that patients who are satisfied with their health provider's communication skills are more likely to adhere to recommendations across the healthcare continuum (Jensen et al. 2011). Poor medication adherence can lead to poor health outcomes, increase in morbidity and mortality rates, as well as, an increase in healthcare costs (Roumie et al. 2011). One strategy for improving medication adherence is through better patient-provider communication. According to Lam et al. nonadherence can be defined as the medication not being taken as prescribed when prescriptions are filled (Lam and Fresco 2015). In developed countries, it is reported that adherence among patients with chronic illnesses averages about 50% (Lam and Fresco 2015). Individuals who do not adhere to their treatment increase the risk of unnecessary health costs or further health complications.

Nevertheless, research has shown that effective patient-provider communication has led to an increase in patient self-efficacy, better self-management behaviors, and compliance to treatment (Roumie et al. 2011). For example, a diabetes study conducted in Northern California investigated the associations between patient communication ratings and cardiometabolic medication refill adherence (Ratanawongsa et al. 2013). The study concluded that through patient-provider communication, medication refill adherence was associated with better cardiometabolic control and reduced morbidity and mortality for individuals diagnosed with diabetes. Furthermore, by utilizing patient-provider communication, doctors were able to engage patients in the health planning process and foster shared decision-making about medications (Ratanawongsa et al. 2013). Good patient-provider communication allows the provider to better identify the health needs of their patients, as well as, better diagnoses of the health condition.

## 10.14   Challenges with Patient-Provider Communication

How health information is communicated to the patient further makes a difference on whether they will acknowledge their health condition and modify their health behaviors. Poor and ineffective communication can lead to misdiagnosis, patient resentment, delayed treatment, or non-adherence to medication (Ha et al. 2010). Miscommunication can also cause complications between the patient and healthcare provider that could hinder patients' understanding, expectations of treatment, or involvement in the medical decisions. Miscommunication further decreases patient satisfaction with the quality of medical care (Ha et al. 2010). A research study conducted on patients seeking outpatient medical care estimated that about 12 million people in the United States are misdiagnosed each year (Firger 2014). Multiple factors influence misdiagnosis including patients having difficulty communicating their symptoms or not revealing all of their family medical history. Other barriers to good patient-provider communication include anxiety and fear, miscommunication, a lack of sensitivity towards the patient, or unrealistic patient expectations (Ha et al. 2010).

The average office visit between a patient and doctor may last about 10–15 min (Togashi 2014). Thus, a doctor is limited by time to properly address a patient's current or chronic medical condition. It also limits the amount of time for the patient to ask follow-up questions, and further limits the sharing of information that could help the patient make a well-informed decision (Togashi 2014). Furthermore, doctors might be dismissive of a patient's explanation or avoid fully discussing a patient's problem if they assume that patient cannot handle the information (Weir 2012). Nondisclosure of information is a potential pitfall that can affect the choices patients make regarding treatment and end-of-life care (Ha et al. 2010).

Another challenge to patient-provider communication is a healthcare provider, who discourages patients from voicing their concerns and expectations and their requests for information. Ha et al. stated that a doctor's discouragement can make a patient feel disempowered and unable to achieve their health goals (Ha et al. 2010). Patients in this situation may tend to resist asking questions or sharing opinions, and again end up less involved in medical decisions (Weir 2012). Moreover, patients may feel intimidated and fear that disagreeing with their provider will lead to negative consequences that could further impact their health care. A patient who takes a passive role in their health care might misinterpret information given by the provider, not care about having a rushed medical visit, and take a less active role in decision making (Epstein et al. 2015).

## 10.15   Strategies and Practice Implications for Improving PCC

Improving patient-provider communication is a continuous process that is necessary to produce better health outcomes. Healthcare providers must be aware of the differences in giving information, as well as, how to encourage patients to be more involved in their health decisions (Verlinde et al. 2012). Providers must also pay close attention to the attitudes that they have towards a patient, which could cause discordance or bias in communication and further impact changes in health behaviors. Nonetheless, the role of patient-provider relationship is characterized by mutual respect, collaboration, as well as, understanding the patient's perspective. Effective patient-provider communication empowers individuals to actively

**Table 10.1** Outcomes of effective communication

| Communication outcomes | Strong patient/family-clinician relationships (trust, rapport, respect, involvement of family and caregivers) |
| | Effective information exchange (recall of information, feeling known and understood) |
| | Validation of emotions (eg, empathy) |
| | Acknowledgement, understanding, and tolerance of uncertainty |
| | Patient participation in decision-making |
| | Coordination of care |
| Intermediate outcomes | Strong therapeutic alliances |
| | Patient knowledgement and understanding |
| | Emotional self-management |
| | High-quality medical decisions (informed by clinical evidence, concordant with patient values, and mutually endorsed) |
| | Family/social support and advocacy |
| | Patient self-efficacy, empowerment, and enablement |
| | Improved adherence, health habits, and self-care |
| Health outcomes | Survival and disease-free survival |
| | Health-related quality of life |
| Societal outcomes | Cost-effective utilization of health services |
| | Reduction of disparities in health and health care |
| | Ethical practice (e.g., informed consent) |

Note: Reprinted from "Patient-centered communication in cancer care: promoting healing and reducing suffering," by Epstein RM, Street RL Jr. Patient-Centered Communication in Cancer Care: Promoting Healing and Reducing Suffering. National Cancer Institute, NIH Publication No. 07-6225. Bethesda, MD, 2007, p. 09

manage their health conditions, adhere to treatment, reduce health disparities, increase patient knowledge, and improve self-efficacy along with other outcomes as shown in Table 10.1 (Brunton 2011; Jenerette and Mayer 2016).

Healthcare providers can use various tools to strengthen patient-provider communication. One way is to increase time for medical visits, which can help advance patient-centered communication, and shared decision-making. According to Verlinde et al. patients have a certain power to control communication during consultation and to influence the provider's communicative behavior (Verlinde et al. 2012). Providers can also encourage patients to prepare a list of questions before coming to their appointment. Having an organized list of questions can facilitate conversations on topics important to the patient (American College of Obstetricians and Gynecologists 2014).

Multiple communication tasks need to be performed by healthcare providers when interacting with their patients. This includes helping patients set appropriate health goals and encouraging more open communication. Providers must also avoid using acronyms and technical words or jargon during visits (Indian Health Service 2016). Moreover, providers must avoid unnecessary details. Keeping explanations simple and clear allows more time for the patient to process the information and formulate questions. On the other hand, patients must be honest about symptoms they experience, as well as, about their family history of any medical conditions. By being authentic and truthful, patients learn to express their concerns and preferences regarding healthcare (Verlinde et al. 2012). Patients will also likely acknowledge their health problems and understand treatment options if they comprehend the information given by the doctor.

Patient-provider communication further requires both the patient and provider to work as a team. Collaborative communication is considered a reciprocal and dynamic relationship that involves the two-way exchange of information (Ha et al. 2010). This type of communication helps doctors with the opportunities to offer and discuss treatment choices to patients and share responsibility with them. Collaborative communication is further improved through communication skills training for providers that may begin in medical school. Although communication skills training has been found to enhance patient-provider communication, regular feedback should be given, as well as, adjustments should be made to develop new skills that will meet the needs of the patient (Ha et al. 2010).

## 10.16   Interventions and Techniques

Various interventions and techniques have been developed to help increase commutation between the patient and healthcare provider. One such technique is known as the Teach-Back Method. The Teach-Back Method is utilized by healthcare providers to check patients understanding of the given health information by asking them to explain in their own words what they need to know or do about their health condition (Agency for Healthcare Research Quality 2014). This technique is not a test of the patient's knowledge, but rather a way to validate their understanding of health information or identify any learning gaps. Research studies have shown that 40–80% of the medical information patients are given during an office visit is forgotten immediately or nearly half of the information retained is incorrect (Agency for Healthcare Research Quality 2014). A patient who lacks clear understanding of their health condition will more than likely not make a follow-up appointment or risk being readmitted to the hospital. Thus, the Teach-Back Method is an effective communication tool to help improve learning outcomes, engage patients and family members in setting health goals, and optimizing health service utilization (Tamura-Lis 2013; Peter et al. 2015). The Teach-Back method minimizes the risk of patients misunderstanding critical information, especially in a medical setting.

Another communication technique used by healthcare providers to empower patients to improve their health outcomes is known as SBAR. SBAR is an acronym that stands of Situation, Background, Assessment, and Recommendation (Dingley et al. 2008; Jenerette and Mayer 2016). The SBAR is considered a collaborative communication strategy that provides structure in patient care situations. The technique is not only effective in bridging the gap in patient-provider communication styles but has also been shown to improve patient safety in healthcare environments (Jenerette and Mayer 2016). For example, a study that focused on nursing home transfers from acute care settings to skilled nursing facilities implemented the SBAR technique. Dingley et al. stated that the use of SBAR helped to avert breakdowns in communication that would affect patients receiving important medications or incomplete information from the nurses (Dingley et al. 2008). Hence, the use of the SBAR technique helps to enhance creditability, trust, and open communication with the provider.

# References

Abel WM, Efird J. The association between trust in health care providers and medication adherence among black women with hypertension. Front Public Health. 2013;1(66):1–6. https://doi.org/10.3389/fpubh.2013.00066.

Agency for Healthcare Research Quality. Using the teach-back technique: a reference guide for health care providers. 2014. http://www.ahrq.gov/professionals/education/curriculum-tools/shareddecisionmaking/tools/tool-6/index.html.

American College of Obstetricians and Gynecologists. Effective patient-physician communication. Obstet Gynecol J Comm Opin. 2014;124(587):389–93.

American Health Information Management Association. What is health literacy? 2016. http://www.myphr.com/HealthLiteracy/

Bailey SC, Oramasionwu CU, Wolf MS. Rethinking adherence: a health literacy–informed model of medication self-management. J Health Commun. 2013;18(1):20–30. https://doi.org/10.1080/108010730.2013.825672.

Barry MJ, Edgman-Levitan S. Shared decision-making—the pinnacle of patient-centered care. N Engl J Med. 2012;366(9):780–1.

Brunton SA. Improving medication adherence in chronic disease management. J Fam Pract. 2011;60(4):S1–8.

Caligtan CA, Carroll DL, Hurley AC, Gersh-Zaremski R, Dykes PC. Bedside information technology to support patient-centered care. Int J Med Inform. 2012;81(7):442–51. https://doi.org/10.1016/j.ijmedinf.2011.12.005.

Center for Disease Control and Prevention. Learn about health literacy. 2015. http://www.cdc.gov/healthliteracy/learn/.

Center for Health Care Strategies. Health literacy fact sheets. 2013. http://www.chcs.org/media/CHCS_Health_Literacy_Fact_Sheets_2013.pdf.

Colbert AM, Sereika SM, Erlen JA. Functional health literacy, medication-taking self-efficacy and adherence to antiretroviral therapy. J Adv Nurs. 2013;69(2):295–304.

Cousin G, Mast MS, Roter DL, Hall JA. Concordance between physician communication style and patient attitudes predicts patient satisfaction. Patient Educ Couns. 2012;87(2):193–7. https://doi.org/10.1016/j.pec.2011.08.004.

Dingley C, Daugherty K, Derieg MK, Persing R. Improving patient safety through provider communication strategy enhancements. 2008. http://www.ahrq.gov/downloads/pub/advances2/vol3/advances-dingley_14.pdf.

Epstein RM, Franks P, Fiscella K, Shields CG, Meldrum SC, Kravitz RL, Duberstein PR. Measuring patient-centered communication in patient–physician consultations: theoretical and practical issues. Soc Sci Med. 2015;61(7):1516–28. https://doi.org/10.1016/j.socscimed.2005.02.001.

Firger J. 12 million Americans misdiagnosed each year. 2014. http://www.cbsnews.com/news/12-million-americans-misdiagnosed-each-year-study-says/.

Gupta A, Sheikh RN. Use emerging digital technology to improve communication with patients. 2014. http://www.kevinmd.com/blog/2014/02/emerging-digital-technology-improve-communication-patients.html.

Ha JF, Anat DS, Longnecker N. Doctor-patient communication: a review. Ochsner J. 2010;10(1):38–43.

Hamine S, Gerth-Guyette E, Faulx D, Green BB, Ginsburg AS. Impact of mHealth chronic disease management on treatment adherence and patient outcomes: a systematic review. J Med Internet Res. 2015;17(2):e52–67. https://doi.org/10.2196/jmir.3951.

Health.gov. Health literacy basics. n.d.. http://health.gov/communication/literacy/quickguide/factsbasic.htm

HealthyPeople.gov. Health communication and health information technology. 2016. https://www.healthypeople.gov/2020/topics-objectives/topic/health-communication-and-health-information-technology.

Indian Health Service. Patient-provider communication toolkit. 2016. https://www.ihs.gov/health-communications/index.cfm?module=dsp_hc_toolkit.

Institute for Healthcare Communication. Impact of communication in healthcare. 2016. http://healthcarecomm.org/about-us/impact-of-communication-in-healthcare/.

Janssen SM, Lagro-Janssen AL. Physician's gender, communication style, patient preferences and patient satisfaction in gynecology and obstetrics: a systematic review. Patient Educ Couns. 2012;89(2):221–6. https://doi.org/10.1016/j.pec.2012.06.034.

Jefferson L, Bloor K, Birks Y, Hewitt C, Blan M. Effect of physicians' gender on communication and consultation length: a systematic review and meta-analysis. J Health Serv Res Policy. 2013;18(4):242–8. https://doi.org/10.1177/1355819613486465.

Jenerette CM, Mayer DK. Patient-provider communication: the rise of patient engagement. Semin Oncol Nurs. 2016;32(2):134–43. https://doi.org/10.1016/j.soncn.2016.02.007.

Jensen F, Gulbrandsen P, DAHL FA, Krupat E, Frankel RM, Finset A. Effectiveness of a short course in clinical communication skills for hospital doctors: results of a crossover randomized controlled trial (ISRCTN22153332). Patient Educ Couns. 2011;84(2):163–9. https://doi.org/10.1016/j/pec.2010.08.028.

Lam WY, Fresco P. Medication adherence measures. BioMed Re Int 2015; ID 217047. https://www.hindawi.com/journals/bmri/2015/217047/. Accessed 01 Feb 2019.

Levinson W, Lesser CS, Epstein RM. Developing physician communication skills for patient-centered care. Health Aff. 2010;29(7):1310–8. https://doi.org/10.1377/hlthaff.2009.0450.

Martin KD, Roter DL, Beach MC, Carson KA, Cooper LA. Physician communication behaviors and trust among black and white patients with hypertension. Med Care. 2013;51(2):151–7.

Nouri SS, Rudd RE. Health literacy in the "oral exchange": an important element of patient–provider communication. Patient Educ Couns. 2015;98(5):565–71.

Peter D, Robinson P, Jordan M, Lawrence S, Casey K, Salas-Lopez D. Reducing readmissions using teach-back: enhancing patient and family education. J Nurs Adm. 2015;45(1):35–42. https://doi.org/10.1097/NNA.0000000000000155.

Ratanawongsa N, Karter AJ, Parker MM, Lyles CR, Heisler M, Moffet HH, Adler N, Warton EM, Schillinger D. Communication and medication refill adherence: the diabetes study of northern California. JAMA Intern Med. 2013;173(3):210–8. https://doi.org/10.1001/jamainternmed.2013.1216.

Rickert J. Patient-centered care: what it means and how to get there. Health Affairs Blog. 2012;24:1–4.

Roumie CL, Greevy R, Wallston KA, Elasy TA, Kaltenbach L, Kotter K, Dittus RS, Speroff T. Patient centered primary care is associated with patient hypertension medication adherence. J Behav Med. 2011;34(4):244–53. https://doi.org/10.1007/s10865-010-9304-6.

Schoenthaler A, Chaplin WF, Allegrante JP, Fernandez S, Diaz-Gloster M, Tobin JN, Ogedegbe G. Provider communication effects medication adherence in hypertensive African Americans. Patient Educ Couns. 2009;75(2):185–91. https://doi.org/10.1016/j.pec.2008.09.018.

Tamura-Lis W. Teach-back for quality education and patient safety. Urol Nurs. 2013;33(6):267–71. https://doi.org/10.7257/1053-816X.2013.33.6.267.

Togashi C. Leading-edge innovations in patient-provider communications. 2014. http://www.healthcareitconnect.com/leading-edge-innovations-in-patient-provder-communications/.

Verlinde E, De Laender N, De Maesschalck S, Deveugele M, Willems S. The social gradient in doctor-patient communication. Int J Equity Health. 2012;11(12):1–14. https://doi.org/10.1186/1475-9276-11-12.

Weir K. Improving patient-physician. Communication. 2012;43(10):36. https://www.apa.org/
    monitor/2012/11/patient-physician.aspx. Accessed 01 Feb 2019.
Ye J, Rust G, Fry-Johnson Y, Strothers H. E-mail in patient–provider communication: a systematic
    review. Patient Educ Couns. 2010;80(2):266–73.
Zill JM, Christalle E, Müller E, Härter M, Dirmaier J, Scholl I. Measurement of physician-
    patient communication—a systematic review. PLoS One. 2014;9(12):e112637. https://doi.
    org/10.1371/journal.pone.0112637.
Zullig LL, Peterson ED, Bosworth HB. Ingredients of successful interventions to improve
    medication adherence. J Am Med Assoc. 2013;310(24):2611–2. https://doi.org/10.1001/
    jama.2013.282818.

# Chapter 11
# Facility

**Olajide Joseph Adebola**

**Abstract**  Healthcare facilities/organizations are places that deal with organization and delivery of healthcare for the general population in a specific location. These facilities/organizations thrive mainly on information collected from patients when they visit them to plan and provide health care. This information collected is known as medical record. Medical record is essential for the provision of quality healthcare by facilities/organizations because it provides the basis for the care received and how it is paid for. The healthcare facilities/organization can only improve the health care given to the patients they serve based on the timely availability of complete medical record. The healthcare providers use the medical record to meet the health-care needs of individual patients. With the arrival of a comprehensive and portable health records, healthcare facilities/organizations can leverage the increased availability of medical records to improve patient care.

**Keywords**  Medical record · Hospital · Portable record · Healthcare facilities Healthcare provider

## 11.1  Introduction

Healthcare facility is also referred to as health facility. In this chapter, healthcare facility is used interchangeably with health organization or health facility, and it refers to a place designated to provide health care to patients (State of New Jersey: Department of Health 2017). Healthcare facilities/organizations are places where health services are planned, organized, and provided to patients. The management of patients takes place in a healthcare facility. Examples of healthcare facility include hospitals, clinics, outpatient care centers, primary healthcare centres, medi-cal camps, burn patient units, feeding centres, etc. (State of New Jersey: Department

O. J. Adebola (✉)
Research and Innovation, Society for Telemedicine and eHealth in Nigeria,
Abuja, FCT, Nigeria

© Springer Nature Switzerland AG 2019
E. R. Onyejekwe et al. (eds.), *Portable Health Records in a Mobile Society*,
Health Informatics, https://doi.org/10.1007/978-3-030-19937-1_11

of Health 2017; World Health Organization: Healthcare Facilities 2018). Healthcare facilities are the bedrock of the national health infrastructure in every national health system. They are the point of health services delivery to the population. The types of healthcare facilities vary from country to country depending on the location, services, and the population served. The location and services provided depend on several factors such as epidemiology of cases that require complex treatment, including accidents; size and density of the population; geographic and climatic conditions; level of economic development; socio-cultural infrastructure; quality and quantity of health resources; national policy for health care; and availability of medical and paramedical personnel (World Health Organization. Regional Office for the Western Pacific 1998). In the United States of America (USA), healthcare facilities can be categorized based on their function, size, location, ownership, and specialization (Gallagher Healthcare 2018). The functions performed by healthcare facilities may include general medical services, specialized services by tertiary facilities (teaching hospitals), care for sudden illnesses (acute care hospitals), management and care for long term conditions (long-term care facilities), healthcare services for the community, ambulatory care services, or research. In terms of size, hospitals can be small size with fewer than one hundred (100) beds, medium size with one hundred to four hundred and ninety-nine (100 to 499) beds, or large size with more than five hundred (500) beds. The location of healthcare facilities may also be used to determine their types. Those in rural areas are referred to as rural healthcare facilities because they serve small communities with limited quantity of healthcare resources and availability of medical and paramedical personnel. The ones in urban areas serve larger populations and have greater health resources and availability of medical and paramedical personnel. Healthcare facilities are classified by the type of ownership, too. Government owned facilities are known as public healthcare facilities and privately owned by corporations or individuals are called private healthcare facilities. There are also healthcare facilities that provide specialized services such as renal centers, that are not tertiary facilities (World Health Organization. Regional Office for the Western Pacific 1998; Gallagher Healthcare 2018). The healthcare facilities must be accessible to patients and meet their various healthcare needs. To have a better understanding of how healthcare facilities /organizations use medical records to improve the care given to the patients they serve, it is necessary to look at some types of healthcare facilities in details.

## 11.2 Types of Healthcare Facilities

The type of healthcare facilities vary from country to country depending on the location, services provided, and the population served. For example, the State of New Jersey Department of Health in USA provides accreditation for the following types of health facilities (State of New Jersey Department of Health: Facility types 2018). They are listed with their definitions.

1. Adult Day Care Services provide preventive, diagnostic, therapeutic, and rehabilitative services under medical and nursing supervision for functionally impaired adult participants. Adult day care service facilities provide services which do not exceed 12 h a day.

2. Alternate Family Care are a contractual arrangement whereby no more than three persons receive room, board, personal care, and other health care services in the home of an unrelated individual who has been approved by a sponsor agency and trained to provide the necessary caregiving.

3. Ambulatory Care Facilities provide preventative, diagnostic, and treatment services to persons who come to the facility to receive services and depart from the facility on the same day.

4. Ambulatory Surgery Centers are surgical facilities in which surgical cases are performed on an outpatient basis. They are licensed as ambulatory surgery facilities and function apart from any other facilities, such as hospitals. (The ambulatory surgery facility may be physically connected to another licensed facility, such as a hospital, but is corporately and administratively distinct.)

5. Assisted Living Programs provide meals and assisted living services, when needed, to residents of publicly subsidized housing which, because of regulations or local housing laws, cannot become licensed as an assisted living residence. An assisted living program may also provide staff resources and other services to a licensed assisted living residence or a licensed comprehensive personal care home.

6. Assisted Living Residences are facilities licensed by the Department of Health to provide apartment-style housing, dining and assisted living services when needed. Apartment units offer, at a minimum, one unfurnished room, a private bathroom, a kitchenette, and a lockable door on the unit entrance.

7. Behavioral Health Management Programs provide specialized long-term care for residents with severe behavior management problems, such as combative, aggressive, and disruptive behaviors.

8. Birth Centers provides routine care to low-risk maternity patients who are expected to deliver babies that are not premature and who will not require surgical intervention.

9. Chronic Hemodialysis Dialysis are services for patients with end stage renal disease where recovery of renal function is not expected.

10. Comprehensive Outpatient Rehabilitation facilities provide comprehensive rehabilitation services to relieve the disabling effects of illness. These include the coordinated delivery of care to maximize the self-sufficiency of the patient.

11. Comprehensive Personal Care Homes provide room and board. They also offer assisted living services when needed. Residential units in comprehensive personal care homes house no more than two residents and have a lockable door.

12. Comprehensive Rehabilitation Hospitals provide comprehensive rehabilitation services to patients to alleviate or ameliorate the disabling effects of illness. These services are characterized by coordinated delivery of care intended to maximize the self-sufficiency of the patient. A rehabilitation hospital can be a stand-alone facility or can be located in a licensed health care facility.

13. Drug Abuse Treatment facilities provide services for methadone detoxification, methadone maintenance, and/or drug-free counseling programs.

14. Family Planning Reproductive health care facilities include services pertaining to contraception, pregnancy detection, options counseling, diagnosis and/or treatment of sexually transmitted diseases, routine gynecological and cancer screening services, health promotion activities, and infertility services. Family planning services may also include prenatal and postpartum care, other gynecological services including colposcopy and cryotherapy, and menopausal services. Family planning services do not include terminations of pregnancies.

15. Family Planning Satellite facilities are affiliates of a separately licensed ambulatory care facility within 30 miles of the licensed ambulatory care facility. The satellite facilities share the same governing authority and provide the same principal services.

16. Federally Qualified Health Centers (FQHC) qualify for enhanced reimbursement from Medicare and Medicaid, as well as other benefits. FQHCs must serve an underserved area or population, offer a sliding fee scale, provide comprehensive services, have an ongoing quality assurance program, and have a governing board of directors.

17. Hemodialysis Dialysis therapy services are provided to either: a hospitalized individual who abruptly sustains loss of kidney function, where dialysis is a temporary life-supporting measure, or in whom recovery of kidney function is expected; or a hospitalized individual with end stage renal disease who requires a regular course of maintenance dialysis therapy.

18. Home Health Agencies provide preventative, rehabilitative, and therapeutic services to patients in a patient's home or place of residence. All home health agencies shall provide nursing, homemaker, home health aide, and physical therapy services.

19. Hospice Branch Hospice Care Programs provide palliative services to terminally ill patients in the patient's home or place of residence, including medical, nursing, social work, volunteer and counseling services.

20. Long-Term Acute Care Hospitals provide acute care through a broad spectrum of clinical care services for acutely ill/medically complex patients who require, on average, a 25 day or longer inpatient stay. They may either be freestanding or a hospital within a hospital.

21. Acute Care Hospitals are facilities for the diagnosis, treatment, or care of individuals suffering from illness, injury, or deformity, and where emergency, outpatient, surgical, obstetrical, convalescent, or other medical and nursing care is rendered for periods exceeding 24 h. Long-Term Care/Subacute Care can be a unit within an acute care hospital.

22. General Hospitals are a unit located within a hospital which utilizes long-term care beds to provide subacute care to patients for a maximum length of stay of 8 days.

23. Long Term Care Pediatric facilities are distinct nursing units, or programs dedicated for occupancy by residents under age 20.

24. Maternal and Child Health Consortiums are voluntarily formed non-profit organizations, incorporated under Section 501(c) (3) of the United States Internal Revenue Code, consisting of all inpatient, ambulatory perinatal, and pediatric care providers and related community organizations in a maternal and child health service region, licensed as central service facility by the Department of Health.

25. Nursing Homes provide health care under medical supervision and continuous nursing care for 24 or more consecutive hours to patients who do not require the degree of care and treatment which a hospital provides, and who, because of their physical or mental condition, require continuous nursing care and services above the level of room and board.

26. Pediatric Day Health Care Services provide additional services in order to provide for the needs of technologically dependent or medically unstable children.

27. Primary Care facilities provide preventive, diagnostic, treatment, management, and reassessment services to individuals with acute or chronic illness. The term is used in reference to facilities providing family practice, general internal medicine, general pediatrics, obstetrics, gynecology, and/or clinical preventive services, including community health centers providing comprehensive primary care.

28. Primary Care Satellites are separately licensed ambulatory care facility, within 30 miles of their parent facility. They share the same governing authority and provide the same principal service as the parent facility.

29. Psychiatric Hospitals are hospitals which provide comprehensive specialized diagnosis, care, treatment and rehabilitation on an inpatient basis for patients with primary psychiatric diagnoses.

30. Residential Health Care Facilities are facilities attached to another licensed long-term care facility, that provide food, shelter, supervised health care and related services in a homelike setting.

31. Satellite Emergency Departments are facilities owned and operated by a hospital, which shall provide emergency care and treatment for patients.

32. Special Hospitals are hospitals which maintain and operate facilities and services for the diagnosis, treatment or care of persons suffering from acute illness, injury or deformity in which comprehensive specialized diagnosis, care, treatment and rehabilitation are administered or performed.

In some countries the list of healthcare facilities could be longer than those of the State of New Jersey. It is important to mention that healthcare facilities play key roles in how modern healthcare is organized and delivered to patients. From the list mentioned above, we see that most healthcare facilities are hospital or hospital based or linked facilities. In order to describe how healthcare facilities use medical record to improve care given to patients, let us consider general hospital services and medical record keeping. A general hospital provides healthcare services for patients of different ages with different diseases. The hospital's main purpose is to

## 11.3    Healthcare Facilities Use Medical Records to Improve Care Given to Patients They Serve

The process of giving services to patients in healthcare facilities starts with documentation of the patient details. Such details become part of the patient's medical record. A medical record is also referred to as health record. Healthcare providers use the health record to document patients' demographics, medical history, findings from medical examinations, diagnosis, and treatment. Detailed and standardized documentation helps tailor the healthcare services available at that facility, as well as facilitate continuity of care for patients. In addition, healthcare facilities use the medical records for medical research and healthcare statistics in order to support the development of a national health system. Medical records must be kept in a standard format in order to facilitate patient care management and organizational reporting. The medical record must be kept safe and always available for use each time patients seek care in a health facility. If the medical record is not available, patients' continuity of care may be compromised, and thus health outcome may be affected (World Health Organization Pacific Region 2002). The medical record consists of four major sections:

1. Administrative data includes demographic and socioeconomic data, such as the name of the patient (identification), sex, date of birth, place of birth, patient's permanent address, and medical record number.
2. Legal data includes a signed consent for treatment by appointed doctors and authorization for the release of information.
3. Financial data relates to the payment of fees for medical services and hospital accommodation.
4. Clinical data includes any clinical findings from medical histories, exams, laboratory and other test results, as well as clinical decisions, treatments, and recommendations for follow up care.

The medical record must be accurate, timely, and accessible by healthcare providers and healthcare facility administrator in order to plan, develop and maintain healthcare services given to patients (World Health Organization Pacific Region 2002; Carpenter et al. 2007; Mann and Williams 2003). The quality of care received by the patient is affected by the quality of the medical record (Adane et al. 2013). Healthcare facilities should use medical records to communicate patients' illnesses and treatment plans among healthcare providers throughout the continuum of care. Medical records are used for research of specific diseases and treatment to improve treatment protocols. Health statistics are collected from the medical record for reporting purposes as well as to improve access to healthcare facilities, expand services at the facilities, and meeting the healthcare needs of patient, overall. Las, hospital services and performance is monitored by reports generated from medical records (World Health Organization Pacific Region 2002; Carpenter et al. 2007; Mann and Williams 2003).

## 11.4 Healthcare Facilities Can Leverage Comprehensive Portable Health Record to Improve Care Given to the Population They Serve

Healthcare facilities can leverage increased availability of comprehensive portable health record to improve care given to the population they serve. Patients visit several healthcare facilities at different stages of their illness and treatment depending on the level of care the facility provides. The accuracy, timeliness, and access to comprehensive patient data available for healthcare provider at the point of care may affect the patient care. Though access is mostly thought to be a challenge related to paper record, it must be noted that electronic health record still face fragmentation because most records exists in electronic health record of the healthcare facility were the record was created.

Comprehensive portable health record is a concept encompassing the patient records that exist on a common, secure, distributed and accessible platform using technologies such as blockchain/distributed ledger used by BitCoin. It is expected that the entry would be validated by the contributor, secured by the platform, and authorized by the patient or their agent for access by healthcare facilities and providers. This would create an opportunity for more efficient and effective delivery of healthcare because it provides timely access to accurate patient health record. In this light, we can look at how healthcare facilities can leverage increase in the availability of comprehensive portable health record through their functions.

1. Provision of Clinical care. With a comprehensive portable health record, the course of illness and treatment of patient at any healthcare facility is available for all healthcare providers involved in the treatment of the patient. Healthcare services available at primary, secondary and tertiary levels of care can be accessed by the patient. Here is an example: a patient that needs Magnetic Resonance Imaging to be done at tertiary hospital will have the report of the test available in a timely manner to share with healthcare providers at another facility. There will be reduction in the turnaround time for follow up consultation and the result can be reviewed without the patient visiting the healthcare facility. A systematic review of the impact of information communication technology (ICTs) on nursing care was conducted in 2015 by Rouleau et al. Their study focused on the nurse practice environment, nursing processes/scope of nursing practices, nurses' professional satisfaction, and patients' outcomes. They reported that ICTs can influence nurses practice environment with impact on quality of care as perceived by nurses (Rouleau et al. 2015).

2. Disease prevention services for the communities. Healthcare facility can use health statistics instantly available from comprehensive portable health record to provide health education and patient engagement communications. Such messages are transmitted to the patients through text messages, instant messages, email communications capabilities between patients and healthcare providers

and healthcare facilities and found to be effective in patient engagement, improve clinical care pathways and health outcomes (Rouleau et al. 2015; Gibbons et al. 2009).

3. Provide advice and support to lower level facilities. This will be facilitated by using telemedicine to provide advice and second opinions for patients directly and to healthcare providers treating the patient. The patient record is readily accessible and the patient can seek a consultation directly from home. Healthcare providers can seek support in managing a difficult case. Bernocchi et al. conducted a non randomised study on home based telemedicine intervention for patients with uncontrolled hypertension in a real life situation. Their findings shows that such telemedicine service provided an opportunity for physician and nurses to review drug treatment through 'telephone linked to remote telemonitoring devices' had a positive effect on blood pressure control of the patients involved in this study. They concluded that the approach decreased the number of hypertensive patients that were currently being treated but who were still uncontrolled (Bernocchi et al. 2014). In the era of comprehensive portable health record, patients' coordination will be easier, faster, and reliable because healthcare providers providing advice and support have instant access to the patient record irrespective of the location of the patient.

4. Provide quality assurance and quality improvement. One of the expected functions of a referral hospital is to develop treatment protocols and standards for disease treatments and train healthcare providers at lower level facilities to use them. Increased availability of comprehensive portable health record can support quality assurance and improvement of healthcare services. Telemedicine was used in a study as an effective intervention to improve antibiotic appropriateness prescription and to reduce costs in pediatric care. The study found that infectious diseases meeting between healthcare facilities and healthcare providers using information communication technology as an economic tool was effective for professional development, and served as multidisciplinary management of complex patients. "It allowed the sharing of protocols and best practices, promoted the prudent use of antibiotics, and it was also associated with direct economic savings. The appropriate use of antibiotics reduces the selection of MDR bacteria and the risk for patients and health workers" (Ceradini et al. 2017). With an increased health record available, developing treatment protocols and sharing them among facilities will be faster and monitoring patient outcome after referrals can be done remotely. The clinical outcomes from such interventions have shown efficacy of deploying telemedicine in chronic disease management (Hersh et al. 2001).

5. Provide education and training. Hospitals are involved in production of healthcare workers in association with universities and medical schools. Online education is readily available today in universities and medical schools. The output research from increased availability of comprehensive health record will enhance the quality of education for healthcare providers to meet the ever increasing demand of quality healthcare services. Medical education via

e-learning platform has provided educational opportunities for students in low- and middle-income countries (Frehywot et al. 2013).

6. Provide management and administrative functions. The capacity of a national health system to monitor and track health output is enhanced with increased availability of portable health record. Patient health records feed directly into national health management information system right at the very time and point of care, irrespective of where the healthcare provider or facility is located. This will enhance planning, coordination, and sharing healthcare resources at the national level.

7. Research and innovation is one of the core functions of a tertiary level hospital. Availability of increased portable health records provides large data to inform research and innovations in healthcare. It will help healthcare providers to improve the processes, methods, and techniques used to provide healthcare in healthcare facilities. Portable health record will allow opportunities of peering among healthcare facilities to collaborate and discover new processes, methods, and techniques that potentially can be used in their own facilities to improve service delivery and patient outcomes (Elrod and Fortenberry 2017). Hannemann-Weber et al. assessed the "contribution of shared communication and decision-making processes in patient-centered healthcare teams to the generation of innovative concepts and consequently to improvements in patient satisfaction" (Hannemann-Weber et al. 2011). Their study looked at the case of patients with rare diseases that regularly face difficulty in accessing expert healthcare provider for treatment. This was coupled with the paucity of experts on such diseases. Treatment of patients with rare disease will require many experts from different healthcare facilities to develop treatment plans that meets the individual needs of such patients. Portable health record will enhance such collaboration. Urquhart et al. did a study in 2012 on "exploring the interpersonal, organization and system level factors that influence the implementation and use of an innovation-synoptic reporting-in cancer care" (Urquhart et al. 2012). They argued favorably on the need for a multidisciplinary team approach in cancer treatment and finding an innovative way to ensure timely decision on treatment plan. Portable health record can fill that gap.

## 11.5   Conclusion

Healthcare facilities in the era of increased available comprehensive portable health record should be able to expand and improve healthcare services given to the population they serve in a timely manner. This will require a lot of re-engineering of clinical care pathway to ensure that the increased availability of health record does not lead to information overload making it difficult for health care provider to make evidence based decision where timing is a critical factor to deliver care.

# References

Adane K, Muluye D, Abebe M. Processing medical data: a systematic review. Arch Public Health. 2013;71:27. https://doi.org/10.1186/0778-7367-71-27.

Bernocchi P, Scalvini S, Bertacchini F, Rivadossi F, Muiesan ML. Home based telemedicine intervention for patients with uncontrolled hypertension: a real life—non-randomized study. BMC Med Inform Decis Mak. 2014;14(52). http://www.biomedcentral.com/1472-6947/14/52

Carpenter I, Ram MB, Croft GP, Williams JG. Medical records and record-keeping standards. Clin Med. 2007;7:328–31.

Ceradini J, Tozzi AE, D'Argenio P, Bernaschi P, Manuri L, Brusco C, et al. Telemedicine as an effective intervention to improve antibiotic appropriateness prescription and to reduce costs in pediatrics. Ital J Pediatr. 2017;43:105. https://doi.org/10.1186/s13052-017-0423-3.

Elrod JK, Fortenberry JL. Peering beyond the walls of healthcare institutions: a catalyst for innovation. BMC Health Serv Res. 2017;17(Suppl 1):402. https://doi.org/10.1186/s12913-017-2342-9.

Frehywot S, Vovides Y, Talib Z, Mikhail N, Ross H, Wohltjen H, et al. E-learning in medical education in resource constrained low- and middle-income countries. Hum Resour Health. 2013;11:4. https://doi.org/10.1186/1478-4491-11-4.

Gallagher Healthcare. Industry insight blog. 2018. https://www.gallaghermalpractice.com/blog/post/what-are-the-different-types-of-hospitals. Accessed 14 Sept 2018.

Gibbons MC, Wilson RF, Samal L, Lehmann CU, Dickersin K, Lehmann HP, Aboumatar H, Finkelstein J, Shelton E, Sharma R, Bass EB. Impact of consumer health informatics applications. Evidence report/technology assessment no. 188. (Prepared by Johns Hopkins University evidence-based practice center under contract no. HHSA 290-2007-10061-I). AHRQ Publication No. 09(10)-E019. Rockville: Agency for Healthcare Research and Quality; 2009.

Hannemann-Weber H, Kessel M, Budych K, Schultz C. Shared communication processes within healthcare teams for rare diseases and their influence on healthcare professionals' innovative behavior and patient satisfaction. Implement Sci. 2011;6:40. https://doi.org/10.1186/1748-5908-6-40.

Hensher M, Price M, Adomakoh S. Referral hospitals. In: Disease control priorities in developing countries. 2006. p. 1230–9. http://www.who.int/management/facility/ReferralDefinitions.pdf. Accessed 15 Sept 2018.

Hersh WR, Helfand M, Wallace J, Kraemer D, Patterson P, Shapiro S, et al. Clinical outcomes resulting from telemedicine interventions: a systematic review. BMC Med Inform Decis Mak. 2001;1:5. https://doi.org/10.1186/1472-6947-1-5.

Mann R, Williams J. Standards in medical record keeping. Clin Med. 2003;3:329–32.

Rouleau G, Gagnon M, Côté J. Impacts of information and communication technologies on nursing care: an overview of systematic reviews (protocol). Syst Rev. 2015;4:75. https://doi.org/10.1186/s13643-015-0062-y.

State of New Jersey Department of Health: Facility types. 2018. https://nj.gov/health/healthfacilities/about-us/facility-types/. Accessed 14 Sept 2018.

State of New Jersey: Department of Health. 2017. https://medlineplus.gov/healthfacilities.html. Accessed 14 Sept 2018.

Urquhart R, Porter GA, Grunfeld E, Sargeant J. Exploring the interpersonal-, organization-, and system-level factors that influence the implementation and use of an innovation-synoptic reporting-in cancer care. Implement Sci. 2012;7:12. https://doi.org/10.1186/1748-5908-7-12.

World Health Organization. WHO guidelines on hand hygiene in health care: first global patient safety challenge clean care is safer care. Geneva: World Health Organization; 2009. Appendix 1, Definitions of health-care settings and other related terms. Available from: https://www.ncbi.nlm.nih.gov/books/NBK144006/. Accessed 17 Sept 2018.

World Health Organization: Healthcare facilities. 2018. http://www.who.int/environmental_health_emergencies/services/en/. Accessed 10 Sept 2018.

World Health Organization Pacific Region. Medical records manual: a guide for developing countries. 2002. http://www.who.int/management/facility/ReferralDefinitions.pdf. Accessed 15 Sept 2018.

World Health Organization. Regional Office for the Western Pacific. District health facilities: guidelines for development and operations. Manila: WHO; 1998. Regional Office for the Western Pacific. http://www.who.int/iris/handle/10665/207020.

# Chapter 12
# Tools

**Olajide Joseph Adebola**

**Abstract** Portable health record provides an opportunity for harvesting large quantities of personal health information of individuals on the go without necessarily visiting a health care facility or health provider. This in turn enables healthcare provider to have access to useful information needed for diagnosis and treatment of patients. Today, more tools are available to healthcare providers when delivering healthcare to individuals. In this context, tools implied are core health informatics technologies, such as software applications, digital medical devices. The use of these tools will greatly improve delivery of health care because of the potential to receive timely personal health information from the individuals' portable health records.

**Keywords** Health record · Health informatics · Technology · Healthcare delivery

## 12.1 Clinical Decision Support

There is a wide variety of tools available at the time and point of care to enhance diagnosis and treatment of individuals. One of such tools, clinical decision support (CDS) software allows the healthcare provider to make better healthcare decisions and act based on dynamic medical knowledge (personal health information), as well as inference mechanisms (usually a set of rules derived from medical experts and evidence based medicine). A clinical decision support software application intelligently uses two or more items of personal health information to generate case-specific advice. Increased availability of patient data from all sources through portable health records along with well-designed CDC software improves decision making at the point of healthcare delivery

O. J. Adebola (✉)
Research and Innovation, Society for Telemedicine and eHealth in Nigeria,
Abuja, FCT, Nigeria

© Springer Nature Switzerland AG 2019                                                      133
E. R. Onyejekwe et al. (eds.), *Portable Health Records in a Mobile Society*,
Health Informatics, https://doi.org/10.1007/978-3-030-19937-1_12

(Healthcare Information and Management Systems Society 2011; Healthcare Information and Management Systems Society 2017).

Here is an example of application of CDS linked to portable health record. An outpatient clinician decides to prescribe a medication that the patient reacted to during the last hospital admission before he or she was discharged. The system generates an alert which warns the clinician about ordering that same medication and it provides alternate medications that can be used to treat the medical condition. At the same time, the system links the clinician to a knowledge resource showing the harmful effects of the drug prescribed originally. The clinician makes an evidenced based decision in providing healthcare to the patient. Such practices are enabled by shared records across healthcare organizations because of the mobility of the health records (Healthcare Information and Management Systems Society 2017).

## 12.2   Electronic Health Record

An Electronic Health Record (EHR) is an electronic storehouse of patient's health information throughout lifetime. It contains details about the health status and health care of an individual and it can be retrieved and used by multiple legitimate users (doctors, nurses and other care givers) in the process of providing healthcare. An EHR allows personal health information of an individual created in hospitals, outpatient practices, and other health care organization to be shared at the time and point of care irrespective of where the individual is seeking healthcare (National Alliance for Health Information Technology 2008; Healthcare Information and Management Systems Society 2015; Tang and McDonald 2006). Access to such information in the EHR prevents unnecessary repetition of diagnostic tests. In addition, the healthcare provider may be alerted when there are changes in the process of diagnosing patients.

## 12.3   Hospital Information Systems

Hospital Information Systems are used by administrators, managers, nurses, doctors or other professionals in the hospital to record administrative information about patient encounters at the hospital. The record contains inpatient admissions, outpatient appointments, attendances, financial information, and laboratory tests (Lærum et al. 2004; Aghazadeh et al. 2012). Portable personal health records have the potential to feed data into hospital information systems, which then can be used by hospital administrators and managers for planning healthcare delivery effectively. Genomic data may also be fed into the hospital information systems but this comes with ethical, legal, and regulatory concerns.

## 12.4   ePrescribing System

Hospital prescribing systems nowadays are using electronic prescribing or ePrescribing system which is the electronically generation and transmission of a medical prescription. ePrescribing is also used to fill and refill prescriptions. ePrescribing is gradually replacing paper and faxed prescriptions in the United States of America and some other countries (Cusack 2008; MedRunner Inc 2011; Salmon and Jiang 2013). With ePrescribing, a portable health record owner is able to receive notification of the electronically transmitted prescription or renew medication authorizations. This tool can also be used for medications delivered to the patient's home.

If the initial pharmacy is closed or does not have the medication on stock, it may be possible to refill the prescription in another pharmacy because the patient controls the record and where it goes, at any time. This makes prescription medications available conveniently and on a timely manner. If there is an error in prescription there is an option to query the source of prescription immediately or stop the pharmacy from delivering the drugs (Cusack 2008; MedRunner Inc 2011; Salmon and Jiang 2013; Genes 2016). The ePrescribing system of the future in an era of portable health records will have the ability to send accurate prescriptions electronically from the healthcare provider to the patient and then to the pharmacy of choice.

## 12.5   Laboratory Information System

Skobelev et al. described laboratory information system as a software application that is able to store and manage all information generated while performing a laboratory analysis (Skobelev et al. 2011). A laboratory system is used to "control and manage samples, standards, test results, reports, laboratory staff, instruments, and work flow automation." Integration of laboratory information management systems with portable health record and the availability of more miniaturized digitized diagnostic devices in microbiology, biochemistry, pathology, hematology or any laboratory discipline will make individual homes to become what may be referred to as "Home Side Laboratory". Basically, the owner of the portable health record conducts self-laboratory testing because the sample is readily available especially in cases of long term medical conditions that require self-monitoring (Skobelev et al. 2011).

These will make it possible to promptly transmit required laboratory data to the healthcare provider and make prompt evidenced-based decisions pertaining to the patient's disease management, particularly related to drug regimens. We will see more mobile laboratories deployed in such fashion, especially, sine the portable health records facilitate health data storage and use for predictive analysis. This makes it easier to maintain the individual health status.

## 12.6    Picture Archiving and Communications (PACS) and Radiology Information System

PACS is a medical imaging technology that allows digital images to be accessed from a wide range of imaging systems or modalities such as ultrasound machine, X-ray, magnetic resonance imaging (MRI) and computed axial tomography (CAT scan) equipment and stored electronically. It is part of radiology information system. It uses a secure network for exchange of patient information which can be viewed from workstations or mobile devices for viewing, processing, and interpreting images. In the era of portable health record, more miniaturized digital imaging technology will be readily available for the patients to use personally and they will determine which healthcare facility or provider interprets the medical image. This technology has already had an impact on the practice of radiology by making more radiologists available through connected healthcare settings when and where there are shortages to offer specialty care.

## 12.7    Mobile Medical Applications

There is an increase in mobile medical application users today including health care providers and patients. This is largely due to the fact that mobile applications help individuals to be involved directly in managing their health and wellbeing. Majority of the existing applications promote healthy living and allow users to have access to health information anywhere anytime, all of which can contribute to individuals becoming active participants in managing their health needs. The tools are readily adopted once developed.

   It is estimated by 2018, 50% of over 3.4 billion smartphone and tablet users will have mobile medical applications on their device. The availability of these applications is the catalyst that will drive the arrival of portable health record where time and space will be insignificant to accessing healthcare delivery. This implies that over 3.4 billion users of smartphone and tablet users will become digital health record officers who will not only be involved in their day to day health plan but determine who and where their health information is used, what it is used for, and how and what health service they want to use (Research 2 Guidance 2015; Food and Drug Administration 2017).

## 12.8    Conclusion

Tools used in the delivery of healthcare integrated with portable health record improve effectiveness in the process of healthcare delivery. There will be substantial changes to the current clinical workflow and patient journey once portable health record become more common. We would have portable health record enabled social interactions just like we use social media tools, such as Twitter or Facebook. Patients will form groups of interest using groupware, messaging systems, and associated team documentation tools, such as Open Notes for patient collaboration, education, and as social support mechanisms. Likewise, the healthcare provider will interact more using such platforms to learn and work as a team.

# References

Aghazadeh S, Aliyev A, Ebrahimnezhad M. Review the role of hospital information systems in medical services development. Int J Comput Theory Eng. 2012;4(6):866.

Cusack CM. Electronic health records and electronic prescribing: promises and pitfalls. Obstet Gynecol Clin N Am. 2008;35(1):63–79.

Food and Drug Administration. FDA mobile medical application. 2017. https://www.fda.gov/MedicalDevices/DigitalHealth/MobileMedicalApplications/default.htm. Accessed 06 Nov 2017.

Genes N. The truth about e-prescribing. 2016. http://epmonthly.com/article/truth-e-prescribing/. Accessed 4 Nov 2017.

Healthcare Information and Management Systems Society. Improving outcomes with clinical decision support. 2011. http://www.himss.org/library/clinical-decision-support. Accessed 06 Nov 2017.

Healthcare Information and Management Systems Society. HER: electronic health record. 2015. http://www.himss.org/library/ehr/. Accessed 17 Mar 2015.

Healthcare Information and Management Systems Society. Scenarios: CDS 101. 2017. http://www.himss.org/library/clinical-decisionsupport/scenarios?navItemNumber=13239. Accessed 01 Nov 2017.

Lærum H, Karlsen TH, Faxvaag A. Use of and attitudes to a hospital information system by medical secretaries, nurses and physicians deprived of the paper-based medical record. 2004. https://doi-org.ezproxy.is.ed.ac.uk/10.1186/1472-6947-4-18. Accessed 04 Nov 2017.

MedRunner Inc. E-prescribing. Inc. 2011. http://www.medrunner.ca/learn-more/eprescribing/. Accessed 5 Nov 2017.

National Alliance for Health Information Technology. Defining key health information technology terms. 2008. http://www.himss.org/content/files/HITTermsFinalReport.pdf. Accessed 04 Nov 2017.

Research 2 Guidance. Mobile health market report 2013–2017. 2015. http://www.research2guidance.com/500m-people-will-be-usinghealthcare-mobile-applications-in-2015. Accessed 06 Nov 2017.

Salmon JW, Jiang R. E-prescribing: history, issues, potential. Online J Public Health Inform. 2013;4(3). https://doi.org/10.5210/ojphi.v4i3.4304.

Skobelev DO, Zaytseva TM, Kozlov AD, Perepelitsa VL, Makarova AS. Laboratory information management systems in the work of the analytic laboratory. Meas Tech. 2011;53(10):1182–9.

Tang PC, McDonald CJ. Electronics health records systems. In: Shortliffe EH, Perreault LE, editors. Biomedical informatics: computers application in healthcare and biomedicine. New York: Springer; 2006. p. 447.

# Chapter 13
# Administration

**Cheryl Austein Casnoff, Roland Gamache, and La Quasha Gaddis**

**Abstract** The purpose of this chapter is to discuss the potential of using health information technologies to help address the current opioid epidemic. The use of technology has become a practical way to disseminate health information. New platforms for integrating medical care and digital technology have been developed to further enhance patient-provider communication. Such technological tools include electronic medical records (EMR), telemedicine, and patient portals. With an ever-growing and rapidly changing healthcare delivery system, it is necessary for providers to adapt to these changes that will help provide an efficient level of care to patients. While individual studies have been published regarding specific technologies, this chapter provides a broad overview of how a range of health IT can serve as tools to prevent and treat opioid abuse. The chapter also addresses some interoperability and privacy challenges that must be addressed to assure the most effective use of health IT to help address the epidemic.

There is general consensus that the overuse of prescription opioids has contributed to the increase in opioid overdose and addiction. To respond to this crisis, it is important to better understand the profile of the abuser of these substances in the community. There are several tools and techniques that have been employed by communities to respond to the opioid crisis.

**Keywords** Opioid · Health IT · Substance abuse · Drug use · Overdose · Telehealth PDMP · Medication assisted treatment · Syndromic surveillance system

---

C. A. Casnoff (✉)
Mitre, Health FFRDC, Windsor Mill, MD, USA

Biological Sciences, Northwestern University, Evanston, IL, USA

Department of Epidemiology and Public Health, Yale Medical School, New Haven, CT, USA

R. Gamache
Gamache Consulting, Chalfont, PA, USA

L. Q. Gaddis
Centers for Disease Control and Prevention, Atlanta, GA, USA
e-mail: laquasha.gaddis@waldenu.edu

## 13.1   Background

Opioids are a class of drugs that have traditionally been used as painkillers, but it is now recognized that they also have great potential for misuse. Drug overdose deaths and opioid-involved deaths continue to increase in the United States and the majority of drug overdose deaths (more than six out of ten) involve an opioid (Centers for Disease Control and Prevention 2017a). According to the Centers for Disease Control and Prevention (CDC), the number of overdose deaths involving opioids (including prescription opioids and heroin quadrupled since 1999 (Centers for Disease Control and Prevention 2017b). In addition, from 2000 to 2015 more than half a million people died from drug overdoses and 91 Americans die every day from an opioid overdose. Deaths from drug overdose are up among both men and women, all races, and adults of nearly all ages (Centers for Disease Control and Prevention 2017b).

This dramatic increase in the use of opioids has been attributed to many causes. First, late in the last century, there was a perception that pain was under-treated, and in 1998 the Joint Commission designated pain as a vital sign. At the same time, drug companies developed and promoted a new generation of synthetic opioids, and doctors prescribed these drugs in increasing quantities. At the same time, illicit forms of opioids became more widely available and abused (Health Affairs 2017).

There is now general consensus that the overuse of these prescription opioids has contributed to the dramatic increase in opioid overdose deaths. For example, the volume of prescription opioids sold to pharmacies, hospitals, and doctors' offices nearly quadrupled from 1999 to 2010. At the same time, deaths from prescription opioids have more than quadrupled since 1999 (Centers for Disease Control and Prevention 2017a). There is growing evidence that while many people benefit from using these medications to manage pain, prescription drugs are frequently diverted for improper use and often lead to addiction. For example, a significant portion of young people who inject heroin report misuse of prescription opioids before starting to use heroin (SAMHSA 2016). Thus, opioid addiction is challenging and a multifaceted public health problem.

The next sections describe emerging evidence about how certain health IT tools can help address this critical epidemic. The way health information is delivered can have a positive impact on healthcare services and addictions, as well as, improve patients' compliance with treatment regimens (HealthyPeople.gov 2016).

## 13.2   Prescription Drug Monitoring Programs (PDMPs)

The growing recognition of the link between prescription opioids and opioid abuse has led to innovative approaches to prescription drug management. One key tool is Prescription Drug Monitoring Programs (PDMPs). PDMPs are not federally operated; they are statewide electronic databases that collect, monitor, and analyze electronically transmitted prescribing and dispensing data submitted by pharmacies and dispensing physicians. Most states have established PDMPs to help providers, including physicians and pharmacists, track controlled substances prescribed to their patients (The Office of the National Coordinator for Health Information Technology 2016).

PDMPs can provide prescribers and pharmacists with key information regarding a patient's prescription history, allowing prescribers to identify patients who are potentially abusing medications (Brady et al. 2014). The data collected usually includes the names and/or demographic information for the patient, prescriber, and dispenser; the name and dosage of the drug; the quantity supplied; the number of authorized refills; and the method of payment (National Council for Prescription Drug Programs 2016). PDMPs are also used by professional licensing boards to identify clinicians with patterns of inappropriate prescribing and dispensing, and to assist law enforcement in cases of controlled substance diversion (Commission on Combating Drug Addiction and the Opioid Crisis Interim Report 2016).

The organization and operation of PDMPs varies among states. Key characteristics include which state agency manages the PDMP, which controlled substances must be reported, how often data are collected and reported, and who can access the PDMP. In some states, the PDMP is "reactive" meaning that only solicited reports are generated in response to a query by authorized users such as prescribers, dispensers and other groups with the appropriate authority. PDMPs in other states can also provide "proactive" or unsolicited reports when there is reason to suspect that violations on the part of the patients or users have occurred but only for law enforcement, not providers (National Council for Prescription Drug Programs 2016).

Key features of PDMPs include (Centers for Disease Control and Prevention 2017c):

**Universal Use** Some states have implemented polices that require providers to check a state PDMP prior to prescribing certain controlled substances and in certain circumstances.

**Real-Time** To date, pharmacists have had flexibility on the frequency of submitting data to PDMPs, ranging from monthly to daily or in "real-time," i.e., under 5 min. Submitting data in real time maximizes the utility of the prescription history data, with significant implications for patient safety and public health.

**Actively Managed** PDMPs can be used by state health departments to help understand the type and locations of the opioid epidemic and to inform and to evaluate interventions alternatives. PDMPs can also be used to send "proactive" reports to authorized users to protect patients at the highest risk and identify inappropriate prescribing trends.

**Easy to Use and Access** Promising practices that states have used to increased use of PDMPs include integrating PDMPs into electronic health record (EHR) systems, permitting physicians to delegate PDMP access to other allied health professionals in their office (e.g., physician assistants and nurse practitioners), and streamlining the process for providers to register with the PDMP.

While the majority of states have created PDMPs, not all states currently require prescribers to utilize the systems. However, in light of the opioid epidemic, more states are now mandating the use of PDMPs by prescribers (Brandeis University 2016). In 2009, for example, Nevada was the first state to require the use of their PDMP if "the practitioner has a reasonable belief that the patient may be seeking the controlled substance, in whole or in part, for any reason other than the treatment of an existing medical condition."(Office of the National Coordinator for Health Information Technology 2012). In 2010, Oklahoma adopted a single substance mandate, requiring that a practitioner consult with the PDMP when prescribing, administering or dispensing methadone. In 2012, Kentucky adopted a more comprehensive mandate with specific criteria defining when all prescribers must utilize the PDMP to view a patient's prescription history. As of May 2016, 30 states had adopted some version of a prescriber use mandate.

Overall, while research on the impact of PDMPs is still somewhat limited, this tool for prescription drug monitoring is viewed as one of the most promising interventions to improve opioid prescribing, inform clinical practice, and protect at risk patients. Early evaluations of PDMPs have shown changes in prescribing behaviors, identification of multiple providers by patients, and decreased substance abuse treatment admissions (Centers for Disease Control and Prevention 2017c). Clinical decision support tools and electronic health record systems that incorporate PDMP data have also shown promise for improving prescribing behaviors and reducing adverse events. As PDMPs continue to be implemented and enhanced across states, their effectiveness and utilization may improve by adding additional tools including collecting data for all controlled substances, proactive reporting to physicians and pharmacists, interstate data sharing, and integration with other health IT systems (Office of the Assistant Secretary for Planning and Evaluation n.d.). As rates of PDMP participation have increased, measures of multiple provider episodes and prescribing of certain drugs have declined. This evidence suggests that PDMP utilization helps to promote medically appropriate prescribing and dispensing and can assist in detecting possible controlled substance misuse and diversion.

## 13.3   Interoperability

State PDMPs vary widely as to whether information is shared with other states. While some states do not allow interstate sharing of information, others have specific practices for sharing (National Council for Prescription Drug Programs 2016). The issue of interoperability, creates challenges to providers, prescribers, and regulatory agencies, and has important implications for the effectiveness of these tools.

PDMPs that link with EHRs must also be easy to use and provide information in a timely manner in order to effectively identify potential drug abuse and diversion, evaluate patient risk, and provide accurate information for clinical decisions at the point of care and within the provider's technology workflow (National Council for Prescription Drug Programs 2016). To address these challenges, the National Council for Prescription Drug Programs (NCPDP) recently developed a set of "Recommendations for an Integrated, Interoperable Solution to Ensure Patient Safe Use of Controlled Substances"(National Council for Prescription Drug Programs 2016). NCPDP recommends that PDMPs be designed to:

- Share real-time information at point of care through the use of existing bidirectional industry standards.
- Reduce the burden on providers by incorporating potential drug abuse information in both pharmacy and prescriber's workflow.
- Enable a proactive notification to providers when a patient exhibits patterns indicative of opioid misuse.
- Ensure access to appropriate therapy for patients with valid medical needs.

In addition, in 2012, the Office of the National Coordinator for Health Information Technology (ONC) in partnership with the Substance Abuse and Mental Health Services Administration (SAMHSA) convened five Work Groups to discuss problems regarding the exchange of PDMP data and developed a set of recommendations to improve the design and use of PDMPs (Office of the National Coordinator for Health Information Technology 2012). The groups recommended that data standards and technical specifications be used for transmitting PDMP data across systems. The groups also developed business-agreement frameworks to help facilitate data sharing. The groups produced policy recommendations to improve PDMP data access and sharing so that dispensers and prescribers can more efficiently and effectively use the information to make real time clinical decisions.

Finally, the recently convened Commission on Combating Drug Addiction and the Opioid Crisis recommended that the federal government provides funding and technical support to states to enhance interstate data sharing among state-based PDMPs to better track patient-specific prescription data and support regional law enforcement in cases of controlled substance diversion. They also recommended that federal health care systems, including Veteran's Hospitals, participate in state-based PDMPs (Commission on Combating Drug Addiction and the Opioid Crisis Interim Report 2016).

## 13.4  Telehealth

In 2014, approximately 21.2 million individuals in the U.S. had a substance abuse disorder (SUD), but only 2.5% of those individuals received treatment (SAMHSA 2017). Widespread increases in opioid addiction and deaths from overdose have led to increasing attention to the lack of access to services in all areas including urban, suburban, and rural areas. While access to treatment is generally lacking in most areas, those areas hit hardest by the increasing rate of drug overdose deaths appear to have the least capacity to provide treatment or services. Furthermore, outpatient SUD treatments and services are four times less likely to be available in rural hospitals than in urban hospitals (HRSA 2016; Medicaid 2018). Persons seeking treatment for drug addiction often have few local options and may have to travel great distances to access services (Brookings 2017).

Early evidence of telehealth use has shown promise as a tool for treating SUDs. A preliminary study that compared a videoconferencing telehealth SUD treatment program with a comparable in-person counterpart from the same organization found that the completion rates were double for the online version compared with traditional outpatient treatment (80% versus 41%, respectively) (National Association of State Alcohol and Drug Abuse Directors 2009). Additional studies suggest that the reasons for increased completion rates using telehealth programs may be convenience and increased confidentiality (National Association of State Alcohol and Drug Abuse Directors 2009; King et al. 2009). Research has also found telehealth services to be as effective as in-person treatment, although small sample sizes may limit the interpretation of the results (SAMHSA 2017).

Regulatory restrictions on prescribing substance use therapy medications (many of which are controlled substances), however, may be constraining providers' willingness to use telehealth for substance use medication treatment (Lacktman 2017). The Ryan Haight Online Pharmacy Consumer Protection Act prohibits controlled substance dispensary through the Internet when no valid prescription has been established during a face-to-face visit or telemedicine visit (National Conference of State Legislatures 2015) The patient must be located in a remote hospital, clinic, or in the presence of another practitioner. Furthermore, variation among state laws on controlled substance-dispensing practices further complicate what can and cannot be provided during a SUD treatment telehealth visit (Center for Connected Health Policy 2018).

One of the first programs designed to help local providers address complex treatment issues like opioid abuse is Project ECHO (Extending Healthcare Community Outcomes). The ECHO model uses telehealth technology to teach local primary care providers specialty skills so they can treat patients themselves, rather than having to refer patients out to services that may require travel to distant sites (Heartland 2017). This type of communication is considered a vital component of patient care. Effective patient-provider communication has been shown to positively influence patients' health behaviors, well-being, and satisfaction with care (Ye et al. 2010). For example, the University of Kansas Center for Telemedicine & Telehealth and Missouri Telehealth Network have been using the ECHO model for chronic pain management which is one of the leading causes of opioid prescribing and abuse. The pain management ECHO teams are promoting alternative tools for pain management to help reduce over-prescribing of opioids. Missouri's ECHO project has also implemented an Opioid Use Disorder ECHO which will focus on decreasing the morbidity and mortality associated with opioid use disorder by giving providers tools to treat these conditions through Medication Assisted Treatment (MAT), which combines behavioral therapy and medication to treat addiction. MAT involves evidence-based care for patients with opioid addictions in primary care offices using both medications and behavioral support to help patients manage their addiction.

Using this model, TelePain is one program that attempts to address geographic barriers to accessing pain specialists for rural populations by connecting physicians at rural community care centers to pain specialists at large medical centers. In partnership with Project ECHO, the TelePain program has been implemented by the University of Washington to increase community providers' knowledge and skills in treating chronic pain (UW Medicine Health System 2017). Specialists offer weekly TelePain sessions, both audio and videoconferencing, for remote community practitioners. The meetings include 30-min presentations on a pain-related topic, followed by time for community providers to present current patient cases for treatment guidance from a multidisciplinary pain team at the University of Washington (Becker's Healthcare 2016; Cummings et al. 2014). Preliminary findings suggest that community physicians were more confident in treating pain cases after participating in the program, patients' self-reported quality of life increased, and opioid doses fell (Doorenbos et al. 2017).

The Agency for Healthcare Research and Quality (AHRQ) is also investing in a series of grants to help support primary care practices and rural communities in delivering MAT through grants totaling approximately $12 million over 3 years (AHRQ n.d.). The goal of these grants is to provide information and tools to local primary care physicians to help them integrate MAT into their practices. The grants are designed to bring together State health departments, academic health centers, researchers, local community organizations, physicians, nurses, and patients in order to offer MAT in local primary care practices.

The grantees will provide access to MAT to over 20,000 individuals struggling with opioid addiction using technology, including patient-controlled smart phone apps, remote training, and expert consultation from Project ECHO. For example, one grantee will support a 23-county project to allow the state's Department of Human Resources, Office of Mental Health and Substance Abuse Services, and the University of Pittsburgh to provide on-site support and training with online services and resources, including telepsychiatry and teleconsults. Medicaid patients with opioid addiction problems across the state will be targeted.

In addition, the U.S. Department of Agriculture (USDA) recently announced the award of $23 million in grants to support 44 distance learning and 36 telemedicine projects in 32 states. These projects are designed to connect rural communities with medical and educational experts across the country, increasing access to health care services, including substance abuse treatment (USDA n.d.). One grantee, for example, will implement a telemedicine system to improve the availability of mental health services in several remote Indiana counties by allowing local health care professionals to connect in real time with urban-based mental health specialists. The program will also connect the urban providers with the rural hospital that serves the community to provide psychiatric diagnoses and support for the hospital emergency department.

## 13.5  Electronic Prescribing

E-Prescribing involves a provider sending a prescription electronically to a pharmacy, such as through an EHR. Electronic prescribing of controlled substances (EPCS) is legal in all 50 states and is designed to help reduce fraud and abuse of controlled substances including prescription opioids. E-Prescribing enables providers to use technology to help avert the problems associated with over prescribing and over use of opioids. Prescribers can be authenticated before prescribing a controlled substance and prescriptions may be transmitted to pharmacies securely without risk of alteration or diversion. As of September 30, 2016, 88.1% of retail pharmacies and 20.2% of e-prescribing providers were enabled for EPCS. Some states, like New York are taking action to mandate electronic prescribing (HealthIT 2018).

## 13.6  Patient Privacy and Information Sharing

Although digital technology has been proven as a useful tool for sending and receiving health information, it can also create a barrier to patient-provider communication. There is growing concern that certain privacy regulations are acting as barriers to communication between providers. This issue is particularly profound as it impacts the opioid epidemic. 42 CFR Part 2 (Part 2), which is a federal regulation governing the confidentiality of drug and alcohol treatment and prevention records and requires addiction treatment professionals to acquire written patient consent before sharing any information with a patient's other health care providers is considered a particular challenge to sharing appropriate health care information (Commission on Combating Drug Addiction and the Opioid Crisis Interim Report 2016). The regulations are designed to provide protections against unauthorized disclosure of substance use records to encourage people with SUDs to seek treatment (Federal Register 2017). While the Part 2 was updated in March 2017 to address issues related to health information exchanges, consent requirements, population health and care management, Part 2 still bars most disclosures of substance abuse treatment information without written consent by the patient and/or his or her personal representative (Department of Health and Human Services 2016).

## 13.7  Decision Support and Data Analytics

The use of geographic information systems (GIS) has long been used by public health to locate geographic clusters of outbreaks to better deploy resources to mitigate the impact of these events on community health. Similar techniques may be used with the opioid issue (Fabrega et al. 1993). However, the opioid crisis is more than just a public health surveillance issue. The crisis concerns pain management (Edlund et al. 2014), particularly participation in daily activities, depression, and other mental health issues (Department of Health and Human Services 2016) including access to mental health services, and the ethical challenges involved in balancing these concerns while providing a thoughtful approach to the opioid crisis. The successful use of GIS to address this issue is not isolated in the United States. For example, Australia has used GIS in a similar fashion to address the over use of opioids in their country as well (Mazumdar et al. 2015).

Social media has become the standard way of communication and is being utlized more increasingly across the health sector to connect to online health communities. Technological innovations have enabled many people to obtain health information, as well as, develop social network groups to share different health experiences (Lefebvre and Bornkessel 2013). Many communities are beginning to use social media to track the assorted issues through the application of technological tools such as Natural Language Processing (NLP) to address and better understand the impact and root cause of this crisis (Conway and O'Connor 2016). This use of social media is also a good source of information to obtain social determinates of health. These social determinates help describe the support network for these patients and are useful for the care team to integrate a care plan that will be supported by the patient's social network.

Fighting an addiction is a difficult task. Some work has begun to use social media to help patients that might be facing a relapse. The information obtained from social media websites can affect how people approach or maintain their health or addictions. It also affects their understanding of how to treat an illness or severe health condition. According to Lefebvre and Bornkessel (2013), health information found online influenced 50% of people to make lifestyle changes, 43% questioned whether to see a doctor, and 38% decided how they would cope with a chronic condition, managed pain or addiction. Systems have been developed to help identify these times of higher risk, not only for substance use, but also for suicide and other mental health issues (McClellan et al. 2017). Many of these times of higher risk are also associated with a period of social isolation, particularly from the patient's close network of support that includes family and close friends.

An area that has shown early promise in identifying substance use disorders is in the use of syndromic surveillance systems in Emergency Departments (EDs) (Goldman-Mellor et al. 2017). Syndromic surveillance uses the chief complaint information as described by the patient when they present for admission to the ED. The chief complaint on admission uses the words of the patient that describes why the patient is seeking care. These complaints are summarized into a list of syndromes using a standard algorithm. An epidemiologic profile is developed from the true cases and this profile is used to develop a set of indicators for this condition that is based on the presenting syndromes (Nolan et al. 2017). The early identification of these patients when they present in the ED will help the physician determine an appropriate treatment for these patients without contributing further long-term burden for the patient.

Similar analytics have been used to also identify ED patients that require substance use treatment by defining a patient profile through the electronic health record (EHR) at the facility (Macias Konstantopoulos et al. 2014). Once again, these decision support systems help providers identify patients that may be physician shopping or requiring other care to obtain a better treatment plan for patient. Other tools for public health use have been developed clustering to look at databases for drug reimbursement. These databases help to identify patients trying to doctor shop (Pauly et al. 2011). One study examined the implantation of an EHR in mental health care. This study provided a great number of improvements in community health outcome (Riahi et al. 2017), but did not address the ethics and legal issues related to patient confidentiality.

A novel area of clinical decision support to aid physicians when prescribing opioids is the use of Fast Healthcare Interoperability Resources (FHIR) as a tool to develop applications that use EHR data and have access to emerging protocols. FHIR follows the HL7 protocols, but does not require the developer to have an intimate knowledge of the HL7 standards. The FHIR platform allows for a form of 'plug and play' development that provides a process to link any new protocols for optimizing pain therapy with the patient information in the EHR (Sinha et al. 2017).

Patient education regarding pain management is an important component for effective treatment. Several physician alert systems, particularly for ED physicians, have in place alternative treatments for patient pain relief prior to the use of opioids. These alerts attempt to select a pain management protocol that will most likely be successful for the patient (Omaki et al. 2017). Additionally, the patient is strongly encouraged to share this information with their primary care provider or their medical home. A strategy to management long-term or chronic pain is important to help prevent other adverse health events including a significant increase in the risk of mortality (Paulozzi et al. 2014; Sims et al. 2007). If chronic pain is considered to be similar to other chronic diseases, then the use of a personal health record (PHR) may be a useful tool for the patient to monitor and contribute to their pain management (Wells et al. 2014). A PHR should be adjusted in such a manner to allow the patient to record data related to their pain symptoms. This record would become a useful tool to determine the effectiveness of different pain treatments, a measure of ability to carry out daily routines with minimum discomfort, and a resource for the provider to make adjustments to the treatment protocols.

Electronic systems have not only been used for public health surveillance and physician alerts. Electronic approaches have also been shown to be effective to prevent abuse in adolescent populations (Hopson et al. 2015). These approaches may have a significant advantage in areas of inadequate access to mental health services, particularly those areas that would require the patient to travel a great distance, such as communities far from urban areas or territories that would not only require travel, but food and lodging as well. This travel also pulls the patient away from the patients local social support network (Jackmon et al. 2016).

## 13.8 Conclusion

As the country continues to grapple with the reality of our opioid epidemic, there are emerging models of prevention, diagnosis and treatment utilizing Health IT that can help empower communities, providers and government stakeholders. Joint efforts will help pursue viable options to take concrete action in addressing the opioid epidemic.

## References

AHRQ. Increasing access to opioid abuse treatment. n.d. https://www.ahrq.gov/professionals/systems/primary-care/increasing-access-to-opioid-abuse-treatment.html. Accessed 14 May 2018.

Becker's Healthcare. The pivotal role of telehealth in value-based care success. 2016. https://www.beckershospitalreview.com/healthcare-information-technology/the-pivotal-role-of-telehealth-in-value-based-care-success.html. Accessed 14 May 2018.

Brady JE, Wunsch H, DiMaggio C, Lang BH, Giglio J, Li G. Prescription drug monitoring and dispensing of prescription opioids. Public Health Rep. 2014;129(2):139–47.

Brandeis University. PDMP prescriber use mandates: characteristics, current status, and outcomes in selected states. 2016. http://www.pdmpassist.org/pdf/COE_documents/Add_to_TTAC/COE%20briefing%20on%20mandates%203rd%20revision.pdf. Accessed 14 May 2018.

Brookings. A nation in overdose peril: pinpointing the most impacted communities and local gaps in care. 2017. https://www.brookings.edu/research/pinpointing-opioid-in-most-impacted-communities/?utm_campaign=Brookings%20Brief&utm_source=hs_email&utm_medium=email&utm_content=56745564. Accessed 14 May 2018.

Center for Connected Health Policy. Online prescribing. 2018. http://www.cchpca.org/online-prescribing-0. Accessed 14 May 2018.

Centers for Disease Control and Prevention. Understanding the epidemic. 2017a. https://www.cdc.gov/drugoverdose/epidemic/index.html. Accessed 14 May 2018.

Centers for Disease Control and Prevention. What states need to know about PDMPs. 2017b. https://www.cdc.gov/drugoverdose/pdmp/states.html. Accessed 14 May 2018.

Centers for Disease Control and Prevention. Opioid overdose. 2017c. https://www.cdc.gov/drugoverdose/pdmp/states.html. Accessed 14 May 2018.

Commission on Combating Drug Addiction and the Opioid Crisis Interim Report. 2016. https://www.whitehouse.gov/sites/whitehouse.gov/files/ondcp/commission-interim-report.pdf. Accessed 14 May 2018.

Conway M, O'Connor D. Social media, big data, and mental health: current advances and ethical implications. Curr Opin Psychol. 2016;9:77–82.

Cummings JR, Wen H, Ko M, Druss BG. Race/ethnicity and geographic access to Medicaid substance use disorder treatment facilities in the United States. JAMA Psychiat. 2014;71(2):190–6.

Department of Health and Human Services. 42 CFR Part 2. 2016. https://s3.amazonaws.com/public-inspection.federalregister.gov/2016-01841.pdf. Accessed 14 May 2018.

Doorenbos A, Eaton L, Theodore B, Sullivan M, Robinson J, Rapp S, et al. (372) TelePain: improving primary care pain management. J Pain. 2017;18(4):S67–S8.

Edlund MJ, Martin BC, Russo JE, DeVries A, Braden JB, Sullivan MD. The role of opioid prescription in incident opioid abuse and dependence among individuals with chronic noncancer pain: the role of opioid prescription. Clin J Pain. 2014;30(7):557–64.

Fabrega H Jr, Ulrich R, Cornelius J. Sociocultural and clinical characteristics of patients with comorbid depressions: a comparison of substance abuse and non-substance abuse diagnoses. Compr Psychiatry. 1993;34(5):312–21.

Federal Register. Confidentiality of substance use disorder patient records. RIN:0930-AA21. Sect. CFR:42 CFR Part 2. 2017. https://www.federalregister.gov/documents/2017/01/18/2017-00742/confidentiality-of-substance-use-disorder-patient-records. Accessed 14 May 2018.

Goldman-Mellor S, Jia Y, Kwan K, Rutledge J. Syndromic surveillance of mental and substance use disorders: a validation study using emergency department chief complaints. Psychiatr Serv. 2017;69(1):55–60.

Health Affairs. Pharmacy benefit management of opioid prescribing: the role of employers and insurers. 2017. https://www.healthaffairs.org/do/10.1377/hblog20170921.062092/full/. Accessed 14 May 2018.

HealthIT. Opioids. 2018. https://www.healthit.gov/playbook/opioid-epidemic-and-health-it/. Accessed 14 May 2018.

HealthyPeople.gov. Health communication and health information technology. 2016. https://www.healthypeople.gov/2020/topics-objectives/topic/health-communication-and-health-information-technology. Accessed 14 May 2018.

Heartland. Opioid epidemic gets a helping hand from telehealth. 2017. http://heartlandtrc.org/opioid-epidemic-helping-hand-telehealth/. Accessed 14 May 2018.

Hopson L, Wodarski J, Tang N. The effectiveness of electronic approaches to substance abuse prevention for adolescents. J Evid Inf Soc Work. 2015;12(3):310–22.

HRSA. Telehealth network grant program. 2016. https://www.hrsa.gov/ruralhealth/programopportunities/fundingopportunities/?id=daf45ff5-c607-43ec-8fcd-5a87b18403a9. Accessed 14 May 2018.

Jackmon W, Blaalid B, Chester-Adam H. Child and adolescent drug abuse prevention: a rural community-based approach. S D Med. 2016:49–54.

King VL, Stoller KB, Kidorf M, Kindbom K, Hursh S, Brady T, et al. Assessing the effectiveness of an Internet-based videoconferencing platform for delivering intensified substance abuse counseling. J Subst Abus Treat. 2009;36(3):331–8.

Lacktman NM. Telemedicine prescribing and controlled substances laws. 2017. https://www.healthcarelawtoday.com/2017/04/03/telemedicine-prescribing-and-controlled-substances-laws/. Accessed 14 May 2018.

Lefebvre RC, Bornkessel AS. Digital social networks and health. Circulation. 2013;127(17):1829–36. https://doi.org/10.1161/CIRCULATIONAHA.112.000897.

Macias Konstantopoulos WL, Dreifuss JA, McDermott KA, Parry BA, Howell ML, Mandler RN, et al. Identifying patients with problematic drug use in the emergency department: results of a multisite study. Ann Emerg Med. 2014;64(5):516–25.

Mazumdar S, McRae IS, Islam MM. How can geographical information systems and spatial analysis inform a response to prescription opioid misuse? A discussion in the context of existing literature. Curr Drug Abuse Rev. 2015;8(2):104–10.

McClellan C, Ali MM, Mutter R, Kroutil L, Landwehr J. Using social media to monitor mental health discussions - evidence from twitter. J Am Med Inform Assoc. 2017;24(3):496–502.

Medicaid. Telemedicine. 2018. https://www.medicaid.gov/medicaid/benefits/telemed/index.html. Accessed 14 May 2018.

National Association of State Alcohol and Drug Abuse Directors. Telehealth in state substance use disorder (SUD) services. 2009. http://nasadad.org/wp content/uploads/2015/03/Telehealth-in-State-Substance-Use-Disorder-SUD-Services-2009.pdf. Accessed 14 May 2018.

National Conference of State Legislatures. Telehealth policy trends and considerations. 2015. https://www.ncsl.org/documents/health/telehealth2015.pdf. Accessed 14 May 2018.

National Council for Prescription Drug Programs. NCPDP's recommendations for an integrated, interoperable solution to ensure patient safe use of controlled substances. 2016. https://www.ncpdp.org/NCPDP/media/pdf/wp/NCPDP_PDMP_WhitePaper_201611-(2).pdf. Accessed 14 May 2018.

Nolan ML, Kunins HV, Lall R, Paone D. Developing syndromic surveillance to monitor and respond to adverse health events related to psychoactive substance use: methods and applications. Public Health Rep. 2017;132(1_suppl):65S–72S.

Office of the Assistant Secretary for Planning and Evaluation. Opioid abuse in the US and HHS actions to address opioid-drug related overdoses and deaths. n.d. https://aspe.hhs.gov/opioid-abuse-us-and-hhs-actions-address-opioid-drug-related-overdoses-and-deaths. Accessed 14 May 2018.

Office of the National Coordinator for Health Information Technology. Enhancing access to prescription drug monitoring programs using health information technology. 2012. https://www.healthit.gov/sites/default/files/work_group_document_integrated_paper_final.pdf. Accessed 14 May 2018.

Omaki E, Castillo R, Eden K, Davis S, McDonald E, Murtaza U, et al. Using m-health tools to reduce the misuse of opioid pain relievers. Inj Prev. 2017. https://doi.org/10.1136/injuryprev-2017-042319

Paulozzi LJ, Mack KA, Hockenberry JM. Vital signs: variation among states in prescribing of opioid pain relievers and benzodiazepines - United States, 2012. MMWR Morb Mortal Wkly Rep. 2014;63(26):563–8.

Pauly V, Frauger E, Pradel V, Rouby F, Berbis J, Natali F, et al. Which indicators can public health authorities use to monitor prescription drug abuse and evaluate the impact of regulatory measures? Controlling high dosage buprenorphine abuse. Drug Alcohol Depend. 2011;113(1):29–36.

Riahi S, Fischler I, Stuckey MI, Klassen PE, Chen J. The value of electronic medical record implementation in mental health care: a case study. JMIR Med Inform. 2017;5(1):e1.

SAMHSA. Opioids. 2016. https://www.samhsa.gov/atod/opioids. Accessed 14 May 2018.

SAMHSA. Behavioral health treatments and services. 2017. https://www.samhsa.gov/treatment. Accessed 14 May 2018.

Sims SA, Snow LA, Porucznik CA. Surveillance of methadone-related adverse drug events using multiple public health data sources. J Biomed Inform. 2007;40(4):382–9.

Sinha S, Jensen M, Mullin S, Elkin PL. Safe opioid prescription: a SMART on FHIR approach to clinical decision support. Online J Public Health Inform. 2017;9(2):e193.

The Office of the National Coordinator for Health Information Technology. 2016 Report to Congress on Health IT Progress. 2016. https://www.healthit.gov/sites/default/files/2016_report_to_congress_on_healthit_progress.pdf. Accessed 14 May 2018.

USDA. USDA Funds 80 Distance Learning and Telemedicine Projects 32 States. n.d. https://www.usda.gov/media/press-releases/2016/07/14/usda-funds-80-distance-learning-and-telemedicine-projects-32-states. Accessed 14 May 2018.

UW Medicine Health System. UW TelePain. 2017. https://depts.washington.edu/anesth/care/pain/telepain/. Accessed 14 May 2018.

Wells S, Rozenblum R, Park A, Dunn M, Bates DW. Personal health records for patients with chronic disease: a major opportunity. Appl Clin Inform. 2014;5(2):416–29.

Ye J, Rust G, Fry-Johnson Y, Strothers H. E-mail in patient–provider communication: a systematic review. Patient Educ Couns. 2010;80(2):266–73.

# Chapter 14
# Technology/Decisions

Egondu R. Onyejekwe

**Abstract** The discussion on portable health records is mostly possible because of the various healthcare technologies that exist today. Implementation of electronic health records, Internet of Things, increased interoperability and exchange of health information, use of the Cloud, development and use of disease management technologies, telemedicine, imaging, and consumer-facing technologies have resulted in major improvement in disease management, improved patient care experiences, and improved patient care outcomes. Artificial intelligence has capitalized on the vast amount of data that is collected on an ongoing basis through the various applications and technologies; thus, improving the potential for further medical research and evidence-driven decision making. New technologies, such as blockchain show promise to delve deeper into new ways of sharing health information among providers and patients. This chapter provides an overview of various healthcare technologies, their current use and potential future use.

**Keywords** Healthcare technology · Portable health records · Artificial intelligence Blockchain · Consumer-facing technologyCloud · Disease management technology · HER · Digital imaging · Internet of things · Interoperability · Telemedicine

## 14.1 Introduction: 10 Top Healthcare Information Technology Trends for 2017

When health data management staff queried various knowledge experts in healthcare IT regarding the type of technologies and trends capable of influencing provider organization for the next 12 months, they were able to compile this list of technologies and trends. The top ten such technology trends on their list included:

E. R. Onyejekwe (✉)
Public Health, Health Administration, College of Health Sciences, Walden University, Minneapolis, MN, USA
e-mail: Egondu.onyejekwe@mail.waldenu.edu

© Springer Nature Switzerland AG 2019    155
E. R. Onyejekwe et al. (eds.), *Portable Health Records in a Mobile Society*, Health Informatics, https://doi.org/10.1007/978-3-030-19937-1_14

1. Artificial Intelligence
2. Blockchain
3. The Cloud
4. Consumer-facing technology
5. Disease Management Technology
6. EHR Improvement
7. Imaging
8. Internet of Things
9. Interoperability
10. Telemedicine

A brief literature review is provided for each of these areas.

## 14.2 Artificial Intelligence

In 2016, there was a surge to apply artificial intelligence (AI) to health. Focus areas included artificial learning, machine learning, and language processing. However, specialized AI, not general AI took the lead for the most tangible applications. Industry experts, predict the emergence and increase of new use cases of specialized AI across verticals and key business processes, which would imply better access to actionable intelligence. To achieve scalable and sustainable AI, vision would, of course, call for good data management (HealthData 2016a, b).

This section begins with a summary of the "Top 12 Ways Artificial Intelligence Will Impact Healthcare" by Jennifer Bresnick (2018) in Health IT Analytics.

Thereafter, the paper by Abby Norman, titled "Diagnosing with "The Stethoscope of the 21st Century"" (Norman 2018).

### 14.2.1 Top 12 Ways Artificial Intelligence Will Impact Healthcare

Bresnick (2018) articulated some top dozen ways Artificial Intelligence (AI) is poised to benefit both providers and users (patients) of the future. AI will become a transformational force in healthcare based upon the following AI-driven tools:

1. *Unifying Mind and Machine Through Brain-Computer Interfaces*
   Patients with neurological diseases and trauma to the nervous system may become incapable of speaking, moving, and interacting meaningfully with others and their environments. Brain-computer interfaces (BCIs) that are backed by AI could restore those fundamental experiences by providing direct interfaces between technology and the human mind in a way that avoids or eliminates the need for keyboards, mice, and monitors.

2. *Developing the Next Generation of Radiology Tools*
   Many of the current diagnostic processes are performed through physical tissue samples from biopsies. These processes are prone to risks including the potential for infection.

   However, certain diagnoses of the inner workings of the human body can be achieved through non-invasive radiological images from MRI machines, CT scanners, and x-rays. Experts predict that the next generation radiology tools are posed to use AI for more accuracy, and in some cases, they will be sophisticated enough to avoid the need for tissue samples. They envision the collaboration of the diagnostic imaging team where the surgeon or interventional radiologist work directly with the pathologist.

3. *Expanding Access to Care in Underserved or Developing Regions*
   *Developing* countries with severe deficits of trained healthcare providers such as ultrasound technicians and radiologists can significantly benefit from AI tools. For example, AI imaging tools can screen, with accuracy equivalent to humans, the chest x-rays for signs of tuberculosis, whose occurrence is high in such places. An AI app with such a capability could be available to providers in low-resource areas. Besides the increased diagnoses and their accuracies, such a tool will reduce the need for an on-site trained diagnostic radiologist. It portends some increase in life expectancy in those areas as well.

4. *Reducing the Burdens of Electronic Health Record Use*
   While electronic health records (EHRs) have advanced the healthcare industry's journey towards digitalization, they have also created a multitude of problems, especially for the providers, such as clinical documentation, order entry, and sorting through the in-basket. Among the problems are therefore, those related to "cognitive overload, endless documentation, and user burnout." Using AI tools such as natural language processing (NLP) for voice recognition and dictation to create more EHR-based intuitive interfaces and automating some of the routine processes, will address some of these short comings. Furthermore, changes like video recording of a clinical encounter, similar to the body cams the police wear can be very useful for healthcare where and if AI and machine learning are used to index those videos for future information retrieval. Additionally, bringing virtual assistants (like the home-use of Siri and Alexa), to the bedside for clinicians to use with embedded intelligence for order entry will prove very beneficial.

5. *Containing the Risks of Antibiotic Resistance*
   The *evolution* of superbugs that are increasingly resistant to antibiotics is a growing concern across the globe. These superbugs that are multi-drug resistant pose difficulties for patients whose infections no longer respond to treatments. These cause many deaths and force patients and health systems to accrue huge financial costs yearly. According to Bresnick (2018) a superbug like "*C. difficile* alone accounts for approximately $5 billion in annual costs for the US healthcare system and claims more than 30,000 lives." With AI tools and machine learning, EHR data can be positioned to identify infection patterns as

well as highlight the patients who could be at risk prior to showing symptoms. Healthcare providers can leverage the machine learning and AI tools to drive the analytics identifying high risk patients thereby enhancing the accuracy of infection identification as well as create faster, more accurate alerts for health-care providers.

6. *Creating More Precise Analytics for Pathology Images*
   *Pathologists* provide one of the most significant sources of diagnostic data for providers across the spectrum of care delivery, says Bresnick (2018) quoting Jeffrey Golden, MD, Chair of the Department of Pathology at BWH and a pro-fessor of pathology at HMS regarding the role of AI in pathology, thus: "'Seventy percent of all decisions in healthcare are based on a pathology result, … Somewhere between 70 and 75 percent of all the data in an EHR are from a pathology result. So, the more accurate we get, and the sooner we get to the right diagnosis, the better we're going to be. That's what digital pathology and AI have the opportunity to deliver.'"

   Finally, on extremely large digital images, those analytics that can drill down to the pixel level and thus enable providers to identify nuances that are not immediately visible to the human eye!

7. *Bringing Intelligence to Medical Devices and Machines*
   The ubiquity of smart devices in the consumer environment, cannot be overem-phasized. In the medical field, they are critical for different uses such as the monitoring of ICU patients. If AI could be used for instance, to enhance the ability to identify deterioration, suggesting that there is a growing sepsis, or purely sense the development of complications, there could be significantly improve outcomes. When negative outcomes are reduced, costs may also be reduced, especially those related to hospital-acquired conditions.

8. *Advancing the Use of Immunotherapy for Cancer Treatment*
   Immunotherapy, especially where the body's own immune system is used to attack malignancies is one of the most promising ways of treating cancers. The problem is that only a handful of patients respond to current immunotherapy options. Worse yet, there is no magic wand that will enable oncologists a pre-cise and reliable method for identifying which patients would be the lucky ones to benefit from this option. Considering machine learning algorithms and their ability to synthesize highly complex datasets, AI shows promise for improving the identification of the lucky patients and may enable the targeting of therapies to particular and unique individual's genetic make-up.

9. *Turning the Electronic Health Record Into a Reliable Risk Predictor*
   While EHRs are a goldmine of patient data, usability issues surface and make their use cumbersome. Data quality and data integrity issues arise when provid-ers and developers are extracting and analyzing that wealth of information, especially where accuracy, timeliness, and reliability are of essence. There is also a mishmash of data formats, structured and unstructured inputs, and incomplete records. These issues make it very difficult to understand exactly how to engage in meaningful risk stratification, predictive analytics, and clinical decision support. Experts are therefore looking at potential AI tools to integrate

the HER data in one place. Other AI tools can also sharpen a provider's understanding of what they are getting when they apply EHR in disease prediction.

10. *Monitoring Health Through Wearables and Personal Devices*

Many of today's consumers have access to devices such as smartphones with step trackers and wearables with sensors capable of collecting valuable data about their health. Tracking a heartbeat, for example, does just that. Used in conjunction with smartphones that are equipped with step trackers, health related data can be captured and analyzed around the clock. Unique individual or general population health data can be gathered when these are supplemented with patient-provided information through apps as well as other home monitoring device/s. The addition of AI apps will play a significant role in extracting actionable insights "from this large and varied treasure trove of data." The challenge will only lie with convincing patients to share data from this intimate and continual monitoring set of devices and applications.

11. *Making Smartphone Selfies Into Powerful Diagnostic Tools*

This potential of harnessing the power of portable devices, can be enriched by taking images with the smartphones and other consumer-grade sources. Thus, clinical quality imaging—will supplement care, especially in underserved populations or developing nations. When such images are available for analysis using AI algorithms, as with dermatology and ophthalmology, the benefits will explode. In the United Kingdom, researchers have developed a tool that identifies developmental diseases by analyzing images of a child's face! (Bresnick 2018). The algorithm is said to be capable of detecting "discrete features, such as a child's jaw line, eye and nose placement, and other attributes that might indicate a craniofacial abnormality." The tool, as of this writing, is capable of providing clinical decision support by matching the ordinary images to more than 90 disorders. As more and more major players in the industry begin and continue to build AI software and hardware into their devices—with the different sensors—there may be more efficient ways to process the more than 2.5 million terabytes of data we generate daily in the digital world. The cell phone manufacturers also believe they can use that data with AI to provide much more personalized and faster and smarter services in healthcare!

12. *Revolutionizing Clinical Decision Making with Artificial Intelligence at the Bedside*

As the healthcare industry gradually shifts away from fee-for-service, it will ultimately shift away from reactive care; thus, advancing every provider's dreams of achieving better care for chronic diseases, reducing costly acute events, and improving sudden deterioration/s of care. The incentivized reimbursement structures allow the industry to develop the processes that will enable proactive and predictive interventions for providers. With AI in the midst, much of the needed bedrock for that evolution will be set. By powering providers with predictive analytics and clinical decision support tools, providers will be clued to problems way before they probably would have recognized them and will have the opportunity to act in timely fashion. For example, conditions such as seizures and sepsis that require highly complex datasets for identification can easily receive earlier warnings from AI. Furthermore, machine learning can also help support provider decisions as well as with the continuation (or none thereof) of care for critically ill patients. Among these are those who have entered a coma after cardiac arrest. Providers will not be forced to visually inspect EEG data from these patients. That method is not only time-consuming, but it is also subjective, which implies that the results may vary based upon the expertise of the individual clinician. Also, an AI algorithm is capable and faster in matching lots of data from many patients, to long term patterns. Furthermore, an AI algorithm could also detect subtle improvements that would impact the clinician's decisions around care.

Consequently, AI provides a more revolutionary approach to data analysis. AI can be leveraged in other promising areas, such as clinical decision support, risk scoring, and early alerting.

When a new generation of tools and systems that are empowered to enable clinicians become more aware of nuances and more efficient when delivering care they more likely to get ahead of developing problems, their efficiency is increased and the cost of care is reduced. The hope is then that AI will "usher in a new era of clinical quality and exciting breakthroughs in patient care."

## 14.2.2   Diagnosing with "The Stethoscope of the 21st Century"

*Turn the page and enter the future—where* 'Your Future Doctor May Not be Human. This Is the Rise of AI in Medicine' as articulated by this Futurism article titled "Diagnosing with "The Stethoscope of the 21st Century"" Abby Norman (2018).

Opined in the article is the view that "Your Future Doctor May Not be Human. This Is the Rise of AI in Medicine." In this rise of AI in medicine, Norman (2018) creates a scenario, where AI—a nameless and a faceless doctor—enters the examination room. It does not present as a robot neither does it present necessarily as the pitting of human minds against machines. Inside that examination room, the role of AI is "to expand, sharpen, and at times ease the mind of the physician so that doctors are able to do the same for their patients." The practicality of this scenario eases the burden for physicians who are already overburdened with clinical and administrative responsibilities. For them, it is also daunting to sort through the massive amount of available information they need to diagnose and treat patients that are provided by the various techniques and tests. That's where having AI as the twenty-first century stethoscope could make all the difference.

So, technology is making its presence known across medical disciplines because the applications for AI in medicine go beyond administrative drudge work. The AI applications range from powerful diagnostic algorithms to finely-tuned surgical robots. Consequently, AI has a definite place in medicine; what is yet to be determined is its value. That value of AI as part of the future contributor of a patient's care team, can be assessed by understanding how AI compares to human doctors. How for instance do AI applications measure up to human doctors with regards to accuracy? Can one articulate the specific, or unique, contributions that AI is able to make? What specific ways can AI be most helpful—and could it be potentially harmful—in the practice of medicine? Answering these questions, will afford confidence to predict, then build, the AI-powered future.

### 14.2.3 AI vs. Human Doctors

A look at the work already done in different places should guide the assessments (Norman 2018). First is diagnosis. While still in the early stages of its development, it is fair to acknowledge that AI is already just as capable as (if not more capable than) doctors in diagnosing patients. For example, AI diagnostics system developed by researchers at the John Radcliffe Hospital in Oxford is reported to be more accurate than doctors at diagnosing heart disease, at least 80% of the time. A "smart" microscope created by researchers at Harvard University, is capable of detecting potentially lethal blood infections. The researchers trained this AI-assisted tool on a series of 100,000 images garnered from 25,000 slides that were treated with dye to make the bacteria more visible. Using the AI system bacteria were already sorted with a 95% accuracy rate. From Showa University in Yokohama, Japan, a different study revealed that a new computer-aided endoscopic system was capable of revealing signs of potentially cancerous growths in the colon "with 94% sensitivity, 79% specificity, and 86% accuracy."

In some cases, researchers are also finding that AI can outperform human physicians in diagnostic challenges that require a quick judgment call, such as determining if a lesion is cancerous.

In 2017, JAMA published a paper by Ehteshami et al. (2017) where AI programs performed better than humans under a time crunch. In this case deep learning algorithms diagnosed metastatic breast cancer better than researches under time pressures. While human radiologists may do well when they have unrestricted time to review cases, timing becomes critical and can make the difference life and death maker in the real world! This is especially true in high-volume, quick-turnaround environments like emergency rooms) where the ability to make a rapid diagnosis becomes prima for saving patients' lives (Ehteshami et al. 2017).

Included in the midst of AI programs by Norman (2018) is IBM's Watson which took barely 10 min to provide actionable advice that took human experts around 160 h. The data was collected from the research that challenged Watson and human experts to glean meaningful insights from the genetic data of tumor cells, and to provide treatment recommendations based on their findings. Also, DeepVariant, a recent Google AI tool that parses genetic data has been proclaimed to be the most accurate tool of its kind (Norman 2018).

Additionally, AI is also a better predictor of health events before they happen, than are humans. According to Norman (2018) researchers from the University of

Nottingham published a study in April 2018 which showed that, AI out-performed the current standard of care in a cardiovascular setting. The self-taught AI trained on extensive data from 378,256 patients, predicted 7.6% more cardiovascular events in patients than the current standard of care. To put that figure in perspective, the researchers indicated also that: "Perhaps most notably, the neural network also had 1.6% fewer "'false alarms"'—cases in which the risk was overestimated, possibly leading to patients having unnecessary procedures or treatments, many of which are very risky."

Big Data and Big Data Analytics are perhaps the most deserving of AI programs. There is a greater need for precision medicine and making sense of huge amounts of data that would be overwhelming to humans. That's exactly what's needed in the growing field of precision medicine. Norman (2018) writes that The Human Diagnosis Project (Human Dx) fulfills that gap of combining machine learning with doctors' real-life experience. The Human Dx is compiling input from 7500 doctors and 500 medical institutions in more than 80 countries to develop a system that anyone—patient, doctor, organization, device developer, or researcher—can access in order to make more informed clinical decisions. That is designing with the end user in mind! Where and when integrated with other health systems, and where people find it useful, they will use it and will not give a second thought to the fact that they are using AI!

For open-minded, forward-thinking clinicians, the immediate appeal of projects like Human Dx is that it would, counterintuitively, allow them to spend less time engaged with technology. "It's been well-documented that over 50% of our time now is in front of a screen," Nundy, who is also a practicing physician in the D.C. area, told Futurism. AI can give doctors some of that time back by allowing them to offload some of the administrative burdens, like documentation.

In this respect, when it comes to healthcare, AI is about augmenting, not replacing doctors, but optimizing and improving their abilities. The derivative of such a system is that it should be both time and place independent. It should be accessible to anyone with a portable device. Anyone can thus integrate it with their portable EHR!

## 14.2.4 Under the (Robotic) Knife

This section ends with Surgical robots (or Robotic knife) AI applications in medicine. A picture is provided in the figure below. The best known is the da Vinci—that functions as an extension of the human surgeon, by controlling the device from a nearby console. Among the more ambitious procedures was one that took place in Montreal in 2010. It was the first in-tandem performance of robotic doctors that included both a surgical robot and a robot anesthesiologist, named McSleepy. The data that was gathered on the procedure reflected the impressive performance of these robotic doctors.

In 2015, MIT conducted a retrospective analysis of FDA data to assess the safety of robotic surgery. During that period, there were 144 patient deaths and 1391 patient injuries reported which were mainly caused by technical difficulties or device malfunctions. The report indicated that while there was a relatively high number of reports, most of the procedures were successful and did not involve any problems. However, the number of events in the more complex cardiothoracic surgery surgical areas were "significantly higher" than in the less complex areas such as gynecology and general surgery.

As of this writing then, while robotic surgery can perform well in some specialties, it is better and safer to leave the more complex surgeries to human surgeons. It is possible that this could change quickly, and surgical robots may become more capable of operating more independently from human surgeons. When such a time arrives, it will surely become harder to apportion blame when something goes wrong.

As a side issue, who would a patient sue for malpractice, robot or the doctor? Litigation is a currently a legal gray area because the technology is very young. Traditionally, though, experts consider medical malpractice as being in the domain of physician negligence or the violation of a defined standard of care. The concept of negligence, therefore begs the question, since by definition, it implies an awareness that AI inherently lacks. However, standards applicable to robots would need to exist if robots could be held to performance standards of some sort.

So, if not the robot, where should the blame lie? Could the human surgeon overseeing the robot be held accountable? Or should the blame be squarely placed on the company that manufactured the robot? Or should it be the specific engineer who designed it? For the present time, there is no clear answer to this problem, but it is worth factoring into the future developments of these programs and their applications.

### 14.2.5   *Mental Healthcare with a Human Touch*

Just a note on an app that addresses human behavioral issues. Cogito is a Boston-based AI and behavioral analytics company that has been using AI-powered voice recognition and analysis to improve customer service interactions in different industries (http://www.cogitocorp.com/). In healthcare, they have Cognito Companion, as a mental health app that tracks a patient's behavior. Cognito Companion monitors a patient's phone for both active and passive behavior signals, including location data that indicates that a patient has been home for many days. It may also collect communication logs that indicate whether the patient has/not texted or spoken on the phone to anyone for several weeks. They indicate that the app only knows if the patient's phone is in use or not for voice or text, and does not track who a user is calling nor what's being said. In any event, the patient's care team can monitor this log and decipher reports for signs that may indicate changes to the patient's overall mental health. Although Cogito now teams up with several healthcare systems in the US to test the app, the veteran population, which is at high risk for social isolation may find this very beneficial. The app acts in a way to build trust and drive engagement in healthcare more broadly. This may be suitable for the veteran population that is reluctant to engage with the healthcare system, especially for mental health resources, and usually because of social stigma. When fortified with AI, the app also uses machine learning algorithms to analyze "audio check ins". These include voice recordings the patient makes, as in an audio diary. The algorithms are also designed to pick up on emotional cues, and match patterns based on energy, intonation, the dynamism or flow in a conversation, similar to what happens when human beings are talking to each other. These are the sources humans use to train the algorithm to learn what "trustworthy" or "competence" sound like, "to identify the voice of someone who is depressed, or the differences in the voice of a bipolar patient when they're manic versus when they're depressed." There are thus simultaneous dual benefits because the app provides real-time information for the patient to track their mood, as well as the information that helps clinicians track their patient's progress over time.

## 14.2.6   Conclusion

It is difficult to conclude this section of the paper on AI because AI is growing, revolutionizing, building, healthcare, but not predicting the future of healthcare. Accenture Consultants opined in a paper titled "Artificial Intelligence (AI): Healthcare's New Nervous System" that AI would be the engine for the growth of healthcare. In healthcare, AI can augment human activity because it is a composite of multiple technologies that enable machines to "sense, comprehend and learn." As a consequence, they can perform both administrative and clinical functions in healthcare, unlike legacy technologies that are simply using algorithms/tools that can only complement a human's activities (https://www.accenture.com/us-en/insight-artificial-intelligence-healthcare).

AI can also unleash immense power through improvements in cost, quality, and access, all of which invariably results in its explosion in popularity. Growth in the AI health market is expected to reach $6.6 billion by 2021—that's a compound annual growth rate of 40%. Accenture Consulting in the same report indicated that the market value of AI in medicine in 2014 was $600 million. But growth in the AI market by 2021, is project to be about $6.6 billion (U.S dollars). That means that the AI health market will grow to more than $10\times^2$ from the current size in merely 3 years from now!

They argue that growth opportunities in healthcare are difficult without significant investment. However, and because artificial intelligence (AI) is a self-running engine, it is more predisposed for growth in healthcare. According to Accenture analysis, key clinical health AI applications, when combined, can by 2026, potentially create $150 billion in annual savings for the US healthcare economy.

By analyzing a comprehensive taxonomy of ten AI applications with the greatest near-term impact in healthcare, Accenture concluded that AI both "thinks and pays for itself" (https://www.accenture.com/us-en/insight-artificial-intelligence-healthcare).

Not only does AI present a significant opportunity for the industry to manage their bottom line in a new payment landscape, but it does enable them to capitalize on new growth potential. To better understand the savings potential of AI, Accenture analyzed a comprehensive taxonomy of ten AI applications with the greatest near-term impact in healthcare. Their assessment yielded three major applications that represent the greatest near-term value. The top three applications are: robot-assisted surgery ($40 billion); virtual nursing assistants ($20 billion); and administrative workflow assistance ($18 billion). The healthcare field is open for these and other AI applications to gain more experience in the field. Because they can also *learn,* their ability to learn and act will invariably lead to improvements in all areas—including "precision, efficiency and outcomes."

In a 2017 article in the Daily Beast by Patel, a case was made for the imperfection of both humans and AI applications. For example, while AI applications do not get tired, they do lack human logic. AI apps can be shown literally thousands or millions of images, without much additional costs for the analysis of each of those images. Furthermore, the AI algorithm will consistently provide the same answer regardless of time, date, and location what it has been trained on. A human dermatologist, on the other hand can mis-diagnose for reasons that range from fatigue to pure volume of images to be analyzed. The errors from such diagnoses can be matters of life and death—yet, there is the other side of the human practitioner that includes feelings, (empathy, love, etc.) that the AI algorithms lack.

But here is another notable difference: Dermatologists have been trained to often use rulers to measure the lesions that they suspect to be cancerous. An AI trained on those biopsy images would most likely say that a lesion was cancerous if a ruler was present in the image! (The Daily Beast 2017).

Therefore, it behooves the currently AI booming health industry not to hurry the integration of AI apps or to do so haphazardly. This is because human logic is not necessarily the same as that of AI/machines. It is also possible that algorithms may also inherit biases for different reasons. Among them are the lack of diversity in the training materials, since AI everywhere trains on the data provided by the researcher and the entity involved. White men, for instance are involved more often in clinical and academic research, because they also dominate these fields and hence data is often biased due to overcollection from white men.

Notable for prior medical decision-making is to emphasized the measure of risk-benefit, to which should be added equality analysis in the future. To that end, the builders of the AI apps should invest more in the effort that is so inclusive that the stakeholders are fully represented in the intended study. Since humans hold the discussions as well as make the decisions about change, they hold the key to how AI is integrated in medicine.

## 14.3  Blockchain

Blockchain technology is a product of the financial industry for cryptocurrency transactions. Interestingly, health and life insurers are becoming engaged and are trying to determine how to adapt blockchain for improving the maintenance of health records, executing health transactions, as well as interact with stakeholders (https://www.healthdatamanagement.com/list/top-healthcare-it-trends#slide-3).

According to, *Abhinav Shashank* (*the CEO & Co-founder at Innovaceer Inc.*) Blockchain developed as the underlying architecture for Bitcoin in 2008 by Satoshi Nakamoto, aka "unknown" or rather a pseudonymous person (or a group). It was used at the time as a core component of the digital currency, 'bitcoin.' Besides the technical jargon, he stated that "blockchain is simply a distributed and a write-once-read-only record of digital events in a chronological order that is shared in a peer-to-peer network." Essentially, blockchain records exchanges and transactions in a shared and distributed database among the entities (users) that are authorized to add records to it, but not allowed to nor able to delete or alter any of the records added up to that point. Furthermore, all users must validate any transaction that occurs.

Soon thereafter, the concept of the blockchain was elevated far beyond simply enabling a decentralized alternative form of currency. Some organizations today use blockchain to apply advanced analytics from distributed sources without compromising the privacy of individuals. That is key for health-based organizations, that are now rapidly applying blockchain-based technology offerings in areas such as healthcare and Pharma/Biotech spaces (Bean and Stephen 2018).

First, what is blockchain? A full discussion of blockchain is provided in a different chapter of this book. But briefly, it is a decentralized and an unconventional platform with a trustless protocol "that combines transparency, immutability, and consensus properties to enable secure, pseudo-anonymous transactions." Consequently, it alleviates the reliance on a single, centralized authority, while still supporting "secure and pseudo-anonymous transactions and agreements directly between interacting parties. It offers decentralization, immutability, and consensus via cryptography and game theory." Because programmable blockchains have generated enormous interest in the healthcare domain, they are regarded as a potential solution for resolving major challenges that range from gapped communications, inefficient clinical report delivery, to fragmented health records (Zang et al. 2017).

Furthermore, there are "smart contracts" (which are codes built on top of block-chains) that can be executed upon predefined conditions. The smart contracts enable the development of Decentralized Apps (DApps). The DApps in turn, can interact with blockchains and support on-chain storage. This IEEE paper assesses blockchain-based DApps in terms of their feasibility, intended capability, and compliance and provides evaluation metrics that will enable the assessment of blockchain-based DApps in the healthcare domain.

At its core, then, blockchain is a distributed system that records and stores transaction records. Specifically, blockchain is a "shared, immutable record of peer-to-peer transactions built from linked transaction blocks and stored in a digital ledger."

Blockchain actually relies on established cryptographic techniques. Consequently each participant—store, exchange, and view information—in a network is allowed to interact without preexisting trust between the parties. In other words, the blockchain system has no central authority, so, transaction records are stored and distributed across all network participants. All participants are known in the blockchain interactions and only require verification by the network before information is added. These enable trustless collaboration between network participants "while recording an immutable audit trail of all interactions."

So, in a nutshell, blockchain is a log of transactions that is replicated and distributed across multiple decentralized locations: As such, 'it offers a secure, high integrity, "neutral" third party mechanism *for knowing what data is where and precisely how it is changing over time*.' It is not a magic bullet that would solve all data management problems, but one that can address those areas that need more improvements in efficiency and security data domains. Among the expected beneficiaries are those of healthcare and Pharma/Biotech (https://ieeexplore.ieee.org/abstract/document/8210842/).

The consulting firm, Deloitte Consulting LLP whose white paper won the Department of Health and Human Services Office of the National Coordinator for Health Information Technology (ONC) sponsored blockchain ideation challenge, presented opportunities for applying blockchain technology to healthcare (Health Information Technology 2016). In their white paper, they articulated how blockchain could make health information exchanges (HIE) more secure, efficient, and interoperable. As a consequence, blockchain can provide a new model for health information exchanges (HIE) by making electronic medical records "more efficient, disintermediated, and secure." While it is not considered a panacea for all HIE needs, it does provide fertile ground for experimentation, investment, and proof-of-concept testing.

They opined that blockchain could facilitate various aspects of health information, such as the "creation of a more comprehensive, secure and interoperable repository of health information." That means that blockchain could drive the elusive move towards "interoperable, comprehensive health records; support smart contracts; help detect fraud; improve provider directory accuracy; simplify the application process; and facilitate a dynamic insurer-client relationship." The most important opportunity is that blockchain technology can transform healthcare—by placing the patient at the center of the healthcare ecosystem, while simultaneously increasing the security, privacy, and interoperability of health data! (https://www2.deloitte.com/us/en/pages/public-sector/articles/blockchain-opportunities-for-health-care.html. Tagged With: Blockchain Technology, Health IT Interoperability, healthcare blockchain).

Blockchain technology provides hope for healthcare by addressing some critical parts of the healthcare ecosystem such as: Ubiquitous, secure network infrastructure; Verifiable identity and authentication of all participants; as well as consistent representation of authorization to access electronic health information, and several other requirements. These are in concert with the nationwide interoperability roadmap provided by the Office of the National Coordinator for Health Information

Technology that defines critical policy and technical components needed for nation-wide interoperability. Because current technologies have limitations related to security, privacy, and full ecosystem interoperability, they are not fully addressing these requirements. Howbeit, although blockchain technology heralds tremendous opportunities for healthcare, it is neither a fully matured current technology nor a panacea that can be readily applied. Among the several challenges blockchain technology faces before complete adoption by healthcare organizations nationwide are several technical, organizational, and behavioral economics challenges!

Current technologies do not fully address these requirements, because they face limitations related to security, privacy, and full ecosystem interoperability.

Blockchain technology presents numerous opportunities for healthcare but it is not fully mature or a panacea that can be immediately applied. Several technical, organizational, and behavioral economics challenges must be addressed before a healthcare blockchain can be adopted by organizations nationwide.

### 14.3.1  Shaping the Blockchain Future

Blockchain technology creates unique opportunities to reduce complexity, enable trustless collaboration, and create secure and immutable information. HHS is right to track this rapidly evolving field to identify trends and sense areas where government support may be needed for the technology to realize its full potential in healthcare. To shape blockchain's future, HHS should consider mapping and convening the blockchain ecosystem, establishing a blockchain framework to coordinate early-adopters, and supporting a consortium for dialogue and discovery.

## 14.4  The Cloud

In 2017, many Healthcare IT executives indicate that the Cloud would receive most of their investments. While many had voiced their reservations with Cloud technology, today, a "HIMSS Analytics Cloud Survey" indicates that more than 83% of healthcare organizations use Cloud technology. Furthermore, "A new report from MarketandMarkets also estimated the healthcare Cloud computing market will grow to $9.48 billion by 2020. The flexibility of Cloud architecture makes it easy to bridge the gaps between the technologies that are already in use at an organization" (https://www.healthdatamanagement.com/list/top-healthcare-it-trends#slide-4).

Cloud computing is very recent, yet, it is a fast-growing area of development in healthcare. The attractions lie with the ubiquitous, on-demand access to virtually limitless or endless resources. This is particularly true when combined with a pay-per-use model that allows for "new ways of developing, delivering and using services." Cloud computing is often used in an "OMICS-context" (Griebel et al. 2015). (OMICS refers biology fields that end in—omics—including genomics, proteomics or metabolomics. Omics aims at the collective characterization and quantification of pools of biological molecules—which invariably translate into form, such as the structure, or a process such as the function, and dynamics of an organism or organisms).

Their chapter on *Exploring the Convergence of Big Data and the Internet of Things*, (Thota et al. 2018) propose an architecture which they claim would be secured, efficient and centralized. Such an architecture is proposed for end to end integration of major health systems. The proposed platform uses Fog Computing to run an end to end framework involvement of an Internet of Things (IoT) based healthcare system that would be deployed in a Cloud environment. Health data would be collected from sensors and securely sent to the near edge devices. The devices, in turn would transfer the data to the Cloud where healthcare professionals would access the data seamlessly. They propose a system that uses asynchronous communication "between the applications and data servers deployed in the Cloud environment."

Because of the security and privacy concerns for patients' electronic health data that are transferred over the Cloud, the authors focus mainly on securing both the authentication and the authorization of all the devices that are crucial for the acceptance and ubiquitous use of IoT in healthcare. There are several aspects of this endeavor and they include identifying and tracking the devices deployed in the system. Additional data is the the location and the tracking of mobile devices, including new devices deployed and connected to existing systems. Also, the location and tracking of apparent communication among the devices and data transfer between remote healthcare systems would be integral parts of the system.

IoT technology enables people and objects to interact with each other in the following areas: smart transport systems; smart cities; smart healthcare; and smart energy. If the healthcare world aggressively pursues the transformations from a location-based system such as a hospital-centered system, to one that is patient or person-centered to then there'll be hope for first, to extend the current healthcare to hospital-home-balanced healthcare in 2020s, and ultimately to home-centered healthcare by the 2030s (Rahmani et al. 2015).

The home-based healthcare would usher in arrangements such as: "human computer interaction, communications, imaging technologies embattled at diagnosis, treatment and monitoring patients without disturbing the quality of lifestyle." Towards that end, it is possible to develop low cost medical devices that can be used for real-time monitoring of patient physical conditions.

The lack of solid security solutions are significant barriers to the adaption of current wireless networks in healthcare. The proposed approaches still have significant issues for IoT-based healthcare applications because of several challenges. Among the challenges are: (1) Medical sensor nodes can be easily lost or abducted because they are tiny in terms of size, (2) Security solutions must be resource-efficient, but medical sensor nodes are limited in processing power, memory, and communication bandwidth.

These resource constraints of medical sensors, make it infeasible to utilize conventional cryptography in IoT-based healthcare (Manogaran et al. 2016b, c, 2017a).

The authors, therefore, proposed the following security protocols: Datagram Transport Layer Security (DTLS) and OpenSSL (OpenSSL is a general-purpose cryptography library. It provides an open source implementation of the Secure Sockets Layer (SSL) and Transport Layer Security (TLS) protocols). To this end, the DTLS handshake protocol is used to provide security solution for the transport layer in IoT; and the Open SSL—an open source project is used for implementing SSL, TLS and various cryptography libraries that include symmetric key, public key, and hash algorithms.

Again, Cloud computing occurs more in "OMICS-context", e.g. for computing in genomics, proteomics and molecular medicine. There are some applications in other healthcare fields. In a 2015 article titled "A scoping review of Cloud computing in healthcare", Griebel et al. analyzed 102 publications on Cloud computing. The aim of their scoping review was to identify in healthcare, both the current state of Cloud computing as well as hot topics in research on Cloud computing that are not part of the traditional domain. The analyses of the 102 publications yielded six main topics that include: telemedicine/teleconsultation; medical imaging; public health and patient self-management; hospital management and information systems; therapy; and secondary use of data. A short brief of why none of these areas actually satisfied the technology/concept of "Cloud computing" is provided below.

### 14.4.1  Telemedicine/Teleconsultation

While most of the articles (34) were about supporting communication and sharing data among stakeholders in healthcare, they did not address Cloud computing. Instead, they mainly described a typical telemedicine application when reporting on the possibility to ubiquitously collect, access and share or analyze patient data, emanating from different hospitals or healthcare providers in dedicated health services networks.

### 14.4.2  Medical Imaging

Medical imaging was another large domain included in 15 of the reviewed articles. The focus was on the storage, sharing, and computation of images. Even the visionary paper in this group, remained on a conceptual level and did not explicitly refer to implementations.

### 14.4.3  Public Health and Patient Self-Management

As with the medical imaging, this area also had 15 articles and covered public health in terms of prevention, health promotion, or improvement for individual citizens and patients. It also covered large population groups (epidemiology). While many of the papers addressed the idea of how Cloud computing might be used to support citizens and patients in managing their health status, they did not differentiate between Clouds and the Internet in general.

### 14.4.4  Hospital Management and Information Systems

There were 13 rather interesting articles that addressed the deployment of clinical information systems into Clouds. From the Commercial HIS vendors to those who chose a more conservative approach by establishing a private Cloud within Seoul National University Bundang Hospital (Korea) none satisfied the requirements for Cloud computing.

### 14.4.5  Therapy

Therapy had seven papers that described applications for planning, managing or assessing therapeutic interventions. All in all, they provided little or no information about Cloud-specific development and some even faced scalability challenges.

### 14.4.6  Secondary Use of Data

The domain of secondary use of data accounted for six papers that stated the use of Cloud computing to enable secondary use of clinical data; e.g. for data analysis, text mining, or clinical research. One of the papers indicated that Cloud computing would offer the advantage of providing researchers with large computing resources. Data security can thus be achieved where/when proprietary Cloud solutions are provided to researchers who can then create their own customized networks and virtual servers. It therefore, remained at the discuss possibilities level. It essentially addressed how to store and share research health data and data from electronic health records in a Cloud structure to reach a HIPAA (Health Insurance Portability and Accountability Act) complying environment. Cloud computing was missing.

### 14.4.7  Conclusion

Cloud computing seem appropriate and a viable solution path towards fulfilling these demands. Consequently, commercial providers like Amazon and Microsoft promise to make hundreds of virtual machines available at ones' fingertips. They posit that they can do so almost immediately and as needed.

As defined by National Institutes of Standards and Technology (NIST), the term "Cloud computing" is "a model for enabling ubiquitous, convenient, on-demand access to a shared pool of configurable computing resources. As essential character-istics of Cloud computing Mell and Grance have listed (1) on demand self-service; (2) broad network access; (3) resource pooling with other tenants; (4) rapid elastic-ity; and (5) measured services. The promise of Cloud computing includes advan-tages "in dynamic resources like computing power or storage capacities, ubiquitous access to resources *at anytime from any place*, and high flexibility and scalability of resources." Therein lies the increased interest for the adoption of Cloud computing in many business areas, including the late adoption in the healthcare domain. There are many articles in healthcare and scientific literature that claim Cloud computing for healthcare applications, which led to this work Griebel et al. (2015).

When reviewing the large amount of most recent literature dealing with Cloud approaches in healthcare it becomes obvious, that many reports are dealing with Cloud-computing technologies as a replacement for grid computing in the OMICS-field, while other fields of application (e.g. health information systems, health information exchange, or image processing and management) still seem to be underrepresented. In the popular literature the application of Cloud computing for healthcare information system provision Cloud computing is often used as a buzz word, but real evidence on research in healthcare Cloud computing (beside the big topic of OMICS) or even its successful and resource saving application is missing. In the biomedical area researchers have proposed Cloud computing as a new busi-ness paradigm for biomedical information sharing.

Broad network access for sharing and accessing data and rapid elasticity to dynamically adapt to computing demands were the features used frequently. Eight of the 102 articles favored the pay-for-use characteristics of Cloud-based services that avoid upfront investments. Twenty-two (22) of the articles presented very gen-eral potentials of Cloud computing in the medical domain, while 66 articles describe conceptual or prototypic projects. So, of the 102 articles, only 14 articles reported from successful implementations.

Furthermore, Cloud computing, in many of the articles was analogized to inter-net-/web-based data sharing. So, the actual characteristics of the particular Cloud computing approach, unfortunately were not presented nor illustrated. They there-fore, caution that Cloud computing could be misrepresented in the healthcare field, either because of the accelerated interest and or in spite of it. Cloud computing in healthcare, for purists can only claim few successful implementations. Yet, many publications claim the term "Cloud" as if it is synonymously for "using virtual machines" or "web-based" while failing to describe the benefit of the Cloud para-digm. They opined that the biggest threats to the adoption of Cloud computing in the healthcare domain are attributable to: the involvement of external Cloud part-ners; and the many unresolved issues of data safety and security. They concluded that Cloud computing is more favorable in these domains: for singular, individual features such as elasticity, pay-per-use; and for broad network access, rather than being upheld as Cloud paradigm on its own!

## 14.5    Consumer-Facing Technology

Consumer-facing technology is rising in popularity. According to West Monroe Partners, a business and technology consulting firm, there is an expansion of digital communication and available patient data which will move the healthcare industry way beyond the conventional doctor's visit. West Monroe Partners found that 91% of the customers in healthcare take advantage of the mobile apps when these are offered. They further stated that 80% of this group prefer mobile to conventional office visit.

Among the other tech initiative are: healthcare insurers, where about 70% of them currently offer rewards programs that harness data from consumer's health tracking devices and apps; and data analytics in healthcare which will advance to help patients in saving money. A critical point at this juncture is to understand and differentiate what is useful data versus data that is simply noise. Useful data will invariably encourage and drive patient action (https://www.healthdatamanagement.com/list/top-healthcare-it-trends#slide-5).

## 14.6    Disease Management Technology

Disease management technology is crucial for healthcare organizations, as they facilitate the speed and shorten the time to market. Healthcare organizations are constantly under pressure not only to differentiate disease treatments, but also to include speed and agility. It is thus, essential to add innovation in mobile, predictive analytics, machine learning and new data driven applications. In totality, these will provide "easily accessible complete views of stakeholders." These views will in turn will enable proper orchestration of customer engagement which would then lead to the achievement of the commercial and R&D goals. Furthermore, the push towards precision medicine, which is believed to have been accelerated in 2017—"with modern data management platforms distilling information down to what really matters for each patient" (https://www.healthdatamanagement.com/list/top-healthcare-it-trends#slide-6).

## 14.7    EHR Improvement

The ubiquity of Electronic Health Records (EHRs) present a catch-22 scenario, according to Brent Lang, CEO of Vocera Communications. EHRs have promoted the digitization of healthcare, but, they have also increased the burden on clinicians. Incessant frustrations among clinicians that result from EHRs range from the lack of interoperability to the excessive documentation burdens placed on them. Brent Lang opined that: "EHR pain points and fatigue are new dynamics; any healthcare technology that addresses those issues will endure" (https://www.healthdataman-agement.com/list/top-healthcare-it-trends#slide-7).

### 14.7.1    Medical Practice Efficiencies and Cost Savings

Many healthcare providers have found that electronic health records (EHRs) help improve medical practice management (HealthIT 2018). This is accomplished by increasing practice efficiencies and cost savings. EHRs benefit medical practices in a variety of ways, including:

- Reduced transcription costs
- Reduced chart pull, storage, and re-filing costs
- Improved documentation and automated coding capabilities
- Reduced medical errors through better access to patient data and error prevention alerts
- Improved patient health/quality of care through better disease management and patient education
- Created more efficient practices

Besides direct patient care, EHR also, enables medical practices reports, such as improved medical practice management. This is especially true when parts are well integrated. Among such parts are the scheduling systems that link appointments directly to progress notes, automate coding, and managed claims. Time is saved with easier centralized chart management, condition-specific queries, and other shortcuts.

Furthermore, there is enhanced communication across the spectrum of providers ranging from other clinicians, labs, and health. The enhanced communication plans run the gamut of:

- Easy access to patient information from anywhere
- Tracking electronic messages to staff, other clinicians, hospitals, labs, etc.
- Automated formulary checks by health plans
- Order and receipt of lab tests and diagnostic images
- Links to public health systems such as registries and communicable disease databases

EHR also affects Revenue in these ways:

- Automating Clinical Documentation and Orders
- Enhanced ability to meet important regulation requirements such as Physician Quality Reporting Initiative (PQRI) through alerts that notify physicians to complete key regulatory data elements
- Reduction of time and resources needed for manual charge entry resulting in more accurate billing and reduction in lost charges

- Reduction in charge lag days and vendor/insurance denials associated with late filing
- Charge review edits alerting physicians if a test can be performed only at a certain frequency
- Alerts that prompt providers to obtain Advance Beneficiary Notice, minimizing claim denials and lost charges related to Medicare procedures performed without Advance Beneficiary Notice

### 14.7.2   Electronic Health Records Reduce Paperwork

EHRs can reduce the amount of time providers spend doing paperwork. Administrative tasks, such as filling out forms and processing billing requests, represent a significant percentage of healthcare costs. EHRs can increase practice efficiencies by streamlining these tasks and significantly decreasing costs.

In addition, EHRs can deliver more information in additional directions. EHRs can be programmed for easy or even automatic delivery of information that needs to be shared with public health agencies or for the purpose of quality measurement.

### 14.7.3   Electronic Prescribing (E-Prescribing)

Paper prescriptions can get lost or misread. With electronic prescribing (e-prescribing), doctors communicate directly with the pharmacy. An e-prescribing system can save lives (by reducing medication errors and checking for drug interactions), lower costs, and improve care. It is more convenient, cheaper for doctors and pharmacies, and safer for patients. In short, e-prescribing is an important, high-visibility component of progress in health information exchange.

### 14.7.4   Electronic Health Records Reduce Duplication of Testing

Because EHRs contain all of a patient's health information in a particular place, providers will be less likely to spend time ordering and searching as well as reviewing patient's information. They are also less prone to duplicate or request unnecessary tests, medical procedures and so forth. Such reduced utilization definitely reduces the overall costs of service/s (HealthIT 2018).

However, unless a patient and the provider remain true to each other and at the same location (city/town/locality), it is hard to incorporate all of a patient's health record in one place. Furthermore, the relationship between quality delivery of care and EHR has not been sufficiently documented.

Jones et al. (2010) conducted a study titled: "Electronic health record adoption and quality improvement in US hospitals." This was published in the American Journal of Managed Care. Their design incorporated national cohort study of from two sources: one consisted of primary survey data about hospital EHR capability collected in 2003 and 2006; and the second source was publicly reported hospital quality data for 2004 and 2007. While these are a little dated, they provide insight towards the objective of the study: "to estimate the relationship between quality improvement and electronic health record (EHR) adoption in US hospitals." They focused on patients with acute myocardial infarction, heart failure, and pneumonia. To assess the relationship between EHR adoption and quality improvement for these conditions, they applied a Difference-in-differences regression analysis.

Their results were quite revealing: the availability of a basic EHR was associated with a significant increase in quality improvement for heart failure (additional improvement, 2.6%; 95% confidence interval [CI], 1.0–4.1%).

However, the adoption of advanced EHR capabilities was associated *with significant decreases* in quality improvement for acute myocardial infarction and heart failure. They "observed 0.9% (95% CI, −1.7% to −0.1%) less improvement for acute myocardial infarction quality scores and 3.0% (95% CI, −5.2% to −0.8%) less improvement for heart failure quality scores among hospitals that newly adopted an advanced EHR, and 1.2% (95% CI, −2.0% to −0.3%) less improvement for acute myocardial infarction quality scores and 2.8% (95% CI, −5.4% to −0.3%) less improvement for heart failure quality scores among hospitals that upgraded their basic EHR."

From these mixed results, they opined that current practices for both the "implementation and use of EHRs have had a limited effect on quality improvement in US hospitals." Of course, potential "ceiling effects" limited the ability of existing measures to assess the effect that EHRs have had on hospital quality, over time. They therefore, proposed the development of "standard criteria for EHR functionality and use," as well as "standard measures of the effect of EHRs on quality…" (Jones et al. 2010).

## 14.8   Imaging

Imaging continues to evolve as it adapts to new reimbursement incentives. Enterprise medical imaging is predicted by Mossis Panner, the CEO of Ambra Health, to eventually eliminate the need for duplicative processing. The Cloud, he opines, will be instrumental in helping physicians solve image management hurdles. Additionally, he stated that value-based care will force the elimination of repeat imaging, as well as enable the broader interoperability. Finally, imaging, will present another important role by helping clinicians in making earlier diagnoses (https://www.healthdata-management.com/list/top-healthcare-it-trends#slide-8).

## 14.8.1  Medical Imaging

Medical imaging involves several different technologies that provide different information about a particular area of the human body that is under study. It is used to diagnose, monitor, or treat medical conditions about body area being studied or treated. These parts could be related to possible disease, injury, or the effectiveness of medical treatment. The different types are presented below.

### Ultrasound Imaging

Ultrasound imaging, also called sonography, applies high-frequency sound waves for viewing the inside of the body. Ultrasound images are captured in real-time, so they are capable of showing both the movement of the body's internal organs and how blood flows through the blood vessels. It does not involve exposures to ionizing radiation as occurs with X-ray imaging.

As shown in the image below, in an ultrasound exam, a transducer (probe) is placed directly on the skin or inside a body opening. The thin layer of sound conducting gel applied to the skin allows the ultrasound waves to be transmitted from the transducer through the gel into the body.

The reflection of the waves off of body structures is used to generate the ultrasound image. The information necessary to produce an image is related to both the amplitude or the strength of the sound signal and the time it takes for the wave to travel through the body.

**Uses**  Physicians use ultrasound imaging as a medical tool to evaluate, diagnose, and treat medical conditions. Common ultrasound imaging procedures include:

- Abdominal ultrasound (to visualize abdominal tissues and organs)
- Bone sonometry (to assess bone fragility)
- Breast ultrasound (to visualize breast tissue)
- Doppler fetal heart rate monitors (to listen to the fetal heart beat)
- Doppler ultrasound (to visualize blood flow through a blood vessel, organs, or other structures)
- Echocardiogram (to view the heart)
- Fetal ultrasound (to view the fetus in pregnancy)
- Ultrasound-guided biopsies (to collect a sample of tissue)
- Ophthalmic ultrasound (to visualize ocular structures
- Ultrasound-guided needle placement (in blood vessels or other tissues of interest)

## MRI (Magnetic Resonance Imaging)

Magnetic Resonance Imaging (MRI) is another medical imaging procedure that produces images of the internal structures of the body, using scanners with strong magnetic fields and radio waves (radiofrequency energy) to generate images. Depending on the part of the body, an MRI scan would last anywhere from 20 to 90 min. An MRI exam works by the creation of an electric current that is passed through coiled wires. These create a temporary magnetic field in a patient's body; radio waves are emitted and sent as well as received by a transmitter/receiver in the machine; digital images of the scanned areas of the body are then produced by these signals. Where contrasts of the MRI image are required, intravenous (IV) drugs like gadolinium-based contrast agents (GBCAs) can also be used.

## Pediatric X-ray Imaging

**X-ray Imaging for Pediatrics** X-ray imaging for pediatrics has improved the diagnosis and treatment of different medical conditions in pediatric patients. The Federal Food, Drug, and Cosmetic Act (FD&C Act), United States Code, Title 21 defines pediatric patients as persons aged 21 or younger at the time of their diagnosis or treatment (https://www.fda.gov/MedicalDevices/ProductsandMedicalProcedures/ucm135104.htm).

Typically, pediatric patients are classified in different groups (neonates, infants, children, and adolescents) based on age ranges. However, and because a patient's size determines the amount of radiation needed to produce a quality medical X-ray imaging, the pediatric patient's size is a more important consideration than age.

Also, while the individual risk from X-ray imaging is small when compared to the benefits that it can provide through helping with accurate diagnosis, it is still important to minimize risk by reducing unnecessary exposure to ionizing radiation because of the following:

- Pediatric patients are more radiosensitive than adults (i.e., the cancer risk per unit dose of ionizing radiation is higher);
- Use of equipment and exposure settings designed for adults may result in excessive radiation exposure if used on smaller patients;
- Pediatric patients have a longer expected lifetime, putting them at higher risk of cancer from the effects of radiation exposure.

Among the medical x-ray imaging exams are computed tomography (CT), fluoroscopy, and conventional X-rays. Consequently, the FDA recommends taking into account the size and age of the patient and that an x-ray image should always be adjusted to meet the needs of the specific type of pediatric patient receiving the exam, and that the equipment should be properly maintained and tested. Furthermore, the FDA recommends that medical x-ray imaging exams should use the lowest radiation dose necessary, whether pediatric patients are grouped by age or by size, and the technique factors used should be based on the clinical indication, patient size, and anatomical area scanned.

## Medical X-ray Imaging

Medical imaging has wide usage in medicine. The many types—or modalities—of medical imaging procedures, present the uses of different technologies and techniques. For example, among those using ionizing radiation to generate images of the body are computed tomography (CT), fluoroscopy, and radiography ("conventional X-ray" including mammography). Ionizing radiation is a form of radiation with enough energy to damage DNA and could potentially cause cancer, by potentially elevating a person's lifetime risk of developing cancer. Appropriately applied medical imaging has led to improvements in the diagnosis and treatment of numerous medical conditions, in both adults and children.

While CT, radiography, and fluoroscopy all work on the same basic principle: these exams differ in their purpose. *Radiography*—such as mammography which is a special type of radiography to image the internal structures of breasts—uses a single image that is recorded for later evaluation. *Fluoroscopy*—is a continuous X-ray image that is displayed on a monitor—allowing for real-time monitoring of a procedure or passage of a contrast agent ("dye") through the body. Care must be applied as fluoroscopy can result in relatively high radiation doses. This is especially true for complex interventional procedures. Among those are placing stents or other devices inside the body. Such procedures may require fluoroscopy to be administered for a long period of time.

Computed tomography (CT) also called "computerized tomography" or "computed axial tomography" (CAT), is a noninvasive medical examination or procedure. It uses specialized X-ray equipment to produce cross-sectional images of the body. A CT exam does involve a higher radiation dose than conventional radiography. Because the CT image is reconstructed from many individual X-ray projections, many X-ray images are recorded as the detector moves around the patient's body. The final products are cross-sectional images or "slices" of internal organs and tissues computer as reconstructed from all the individual images. CT scans can be performed on every region of the body for a variety of reasons (e.g., diagnostic, treatment planning, interventional, or screening). Most CT scans are performed as outpatient procedures.

Of course, while X-rays and CT exams represent major advances in medicine there are also risks amongst the many benefits associated with their application.

X-ray imaging exam is a valuable medical tool for a wide variety of examinations and procedures that are used for: noninvasively and painlessly help to diagnosis disease and monitor therapy; support medical and surgical treatment planning; and guide medical personnel as they insert catheters, stents, or other devices inside the body, treat tumors, or remove blood clots or other blockages.

CT images of internal organs, bones, soft tissue, and blood vessels provide greater clarity and more details than conventional X-ray images, such as a chest X-Ray.

### 14.8.2   Benefits/Risks

The benefits of a CT scan far exceed the risks when used appropriately. CT is a valuable medical tool that helps a physician with: the diagnosis disease, trauma or abnormality; the planning of treatment and guiding interventional or therapeutic procedures and; the monitoring of the effectiveness of therapy (e.g., cancer treatment). For some occasions, detailed CT images may eliminate the need for exploratory surgery.

A major concern about CT scans include the risks from exposure to ionizing radiation which may cause a small increase in a person's lifetime risk of developing cancer. This type of exposure is of particular concern in pediatric patients since the cancer risk per unit dose of ionizing radiation is higher for younger patients than adults. Furthermore, younger patients have a longer lifetime for the effects of radiation exposure to manifest as cancer.

A second concern entails possible reactions to the intravenous contrast agent, or dye, which may be used to improve visualization.

Howbeit, the risk from a medically necessary imaging exam is relatively insignificant in both children and adults, especially when compared to the benefit of accurate diagnosis or intervention in both groups. For children especially, it is very important to make sure that CT scans are performed with appropriate exposure factors.

## 14.9   Internet of Things

The Internet of things (IoT) is expected to increase with the growing ability to connect medical devices and other gadgets to the Internet. With this growth comes the security challenges. For example, an IoT recently facilitated a major distributed denial of service attack that resulted in major organization outages. As IoT grows, this problem will also grow, unless networks and connected devices are adequately safeguarded (https://www.healthdatamanagement.com/list/top-healthcare-it-trends#slide-9).

Islam et al. (2015) describe the Internet of Things (IoT) thus:

"The Internet of Things (IoT) is a concept reflecting a connected set of anyone, anything, anytime, anyplace, any service, and any network. The IoT is a megatrend in next-generation technologies that can impact the whole business spectrum and can be thought of as the interconnection of uniquely identifiable smart objects and devices within today's internet infrastructure with extended benefits." The IoT provides appropriate solutions for application in a variety of areas such as smart cities, traffic congestion, waste management, structural health, security, emergency services, logistics, retails, industrial control, and healthcare. So, it is plausible to introduce IoT in every field. Their benefits accrue typically, from the advanced connectivity of the associated devices, systems, and services.

Both medical care and healthcare represent several attractive areas for the IoT applications including remote health monitoring, fitness programs, chronic diseases, and elderly care, treatment and medication compliance both at home and by healthcare providers. Consequently, the different medical devices, sensors, and diagnostic and imaging devices can become the needed smart devices or objects that constitute a core part of the IoT.

Given the above, IoT-based healthcare services are "expected to reduce costs, increase the quality of life, and enrich the user's experience." For the healthcare providers, and using remote provisions, the IoT has the potential to reduce device downtime, correctly identify optimum times for replenishing supplies for various devices so they can run smoothly and continuously as well as provide for the efficient scheduling of limited resources, especially in resource and skill poor areas. Thus, the "IoT revolution is redesigning modern healthcare with promising technological, economic, and social prospects" (Islam et al. 2015).

While technology can never replace humans as the ultimate decision maker, Shashank (2017) indicated that Internet of Things is a game-changer and provided three words that define the Internet of Things (IoT) as: "Convenience. Efficiency. Automation." IoT involves technology that attempts to connect different devices with on and off switches to the internet. IoT then, captures and monitors data on devices that are connected to the Cloud. Consequently, IoT should and does impact healthcare. Shashank (2017) provided six reasons why Healthcare needs IoT. These are:

1. To turn data into actions—since health will be quantifiable, it will be useful to take advantage of quantified health technology as this would enable better object measurement and tracking of health for better outcomes, that would increase healthcare performance.
2. To improve patient Health—because of the potential to assess the proper functioning of wearable health devices on patients, such as heart monitors, the immediacy of updating that information on the Cloud and also sharing with other devices would deter the slow process of fetching the electronic health records. Because tiny details are captured in the process, more precise and advantageous decisions are made for the patient. It can also be used as a medical adherence and home monitoring tool.
3. To Promote Preventive Care—IoT provides ubiquitous access to real-time, high fidelity data on a given individual's health that is pertinent for the management of the growing area of preventive care. It will help people in both the prevention of diseases and in living healthier lives.
4. To Enhance Patient Satisfaction and Engagement—in areas such as the optimization of surgical workflow, as in informing about a patient's surgical discharge

from surgery to their families. Devices connected directly to the Internet are more prone to deliver valuable data, that can accelerate patient engagement with physicians, rather than engaging in direct patient-physician interaction.

5. To advance Care management—wearable for heart rate, sleep, perspiration, temperature and activity allow different care teams to collect millions of data points on personal fitness. Furthermore, the sensor-fed information can release alerts to both patients and providers in real-time. These attributes augur well for extensive workflow optimization, which invariably ensure the management of care from home.

6. To advance Population Care Management—the ability of providers to receive insight driven optimization and use IoT for both home monitoring and chronic disease management makes them vital. IoT has the ability to integrate devices and observe the growth of wearables, which are easier to use in detecting the data that is missing in the EHR.

It is beyond imagination as well as mind-boggling to conceive what the future of healthcare can be if IoT was expanded and invested in. For example, imagine a patient who was notified about his visit to his doctor who is also simultaneously informed, simply because their calendars were connected through IoT. On his way, the patient encounters some traffic delay and his car is able to send a text to the doctor about the impending delay. This, sure, is a technology that will optimize every aspect of care and transform the way care is managed.

Shashank (2017) articulated the barriers towards the adoption of IoT in healthcare, and cites the more obvious challenges as storing, managing and securing data as well as the persistent problem regarding lack of EHR integration. Additionally, he stated that reliability and security issues with data as well as interoperability present problems. Finally, a lack of training and poor infrastructure present problems, because even where and when data flows freely, many providers lack both the infrastructure and the know-how to access it.

Besides the problems at the back-end (provider/care giver end) there are those associated with the front end—the populations that can benefit most from IoT. This population of beneficiaries may beset with poor internet access—including vulnerable populations such as the elderly, those with low education levels, lower-income populations, as well as rural residents and minorities.

Industries like automobile, industrial, civil planning, and retail are experiencing the popularity of IoT and are seeing it soar. In these areas, connected devices talk to each other and smoothen operations. IoT can do the same in healthcare and solve a myriad of problems by helping optimize the way things are done. Among the benefits of such connected technologies are providers who will observe fewer missed appointments; improved adherence to care plans, and improved outcomes such as reduced inpatient admissions. When and if fully adopted, IoT will align the shared goals in healthcare that include: better health, reduced costs, and improved experience of all healthcare stakeholders—from patients to providers and care givers! (Shashank 2017).

## 14.10   Interoperability

Interoperability is at the core of moving electronic health records (EHRs). The 21st Century Cures Act makes provisions for the EHRs. A provision in the Act seeks to improve interoperability, thereby improving patient care, and thus, more effort is expected in the information exchange initiatives as well as in the development and use of FHIR (https://www.healthdatamanagement.com/list/top-healthcare-it-trends#slide-10).

FHIR, is an HL7 emerging information exchange standard and stands for 'Fast Healthcare Interoperability Resources' Specification, which is a standard for exchanging healthcare information electronically. Such specifications are necessary because healthcare records are becoming increasingly digitized.

Furthermore, and as patients move through the healthcare ecosystem, it is imperative that "their electronic health records become available, discoverable and understandable." In addition, the data must be structured and standardized to support machine-based processing and systems such as the clinical decision support. To be of value, there must be interoperability that allows data to move across systems. Interoperability depends on standards.

FHIR is one such standard, but offers more than a standard that solves existing problems in interoperability, for besides interoperability, it also provides a platform for the future. It galvanizes the healthcare filed to the impending healthcare transformation that is heralded by the convergence of biological and information revolutions, by economic imperative and by social change. FHIR's strength lies in it being grounded in the real world which is changing rapidly. Interoperability is "'all about the people'; to get past the peak of inflated expectations to the plateau of productivity." The impending high-level disruption demands a shared vision of the future that helps place FHIR into a wider context.

HL7 has been in the forefront of addressing the movement of data, including the production of healthcare data exchange and information and standards for information modeling for over 20 years. Today, FHIR can be used as a stand-alone data exchange standard but can also be used in partnership with existing widely used standards. FHIR has thus, emerged as a new specification that is based on emerging industry approaches, as well as industry that has informed HL7 through years of lessons around requirements, successes and challenges gained through defining standards for the interoperability in the healthcare industry (https://www.hl7.org/fhir/overview.html).

### 14.10.1   What Is Interoperability?

"Interoperability describes the extent to which systems and devices can exchange data, and interpret that shared data. For two systems to be interoperable, they must be able to exchange data and subsequently present that data such that it can be understood by a user" (http://www.himss.org/library/interoperability-standards/what-is-interoperability).

The Interoperability Imperative is that Value-Based Care ultimately Depends on Health Information Exchange, which inevitably depends on Interoperability (HIMSS17 Infocus 2017).

HIMSS (Health Information Management Systems Society) defines interoperability and *The HIMSS Board approved the following definition of interoperability on April 5, 2013*: In healthcare, interoperability is the ability of different information technology systems and software applications to communicate, exchange data, and use the information that has been exchanged (HIMSS 2010).

American Academy of Family Physicians (AAFP), Center for Health IT (2013) holds: Data exchange schema and standards should permit data to be shared across clinician, lab, hospital, pharmacy, and patient regardless of the application or application vendor.

Interoperability means the ability of health information systems to work together within and across organizational boundaries in order to advance the effective delivery of healthcare for individuals and communities (HIMSS 2013).

There are three levels of health information technology interoperability provided by NCVHS (2000). These include:

- "*Foundational*" interoperability allows data exchange from one information technology system to be received by another and does not require the ability for the receiving information technology system to interpret the data.
- "*Structural*" interoperability is an intermediate level that defines the structure or format of data exchange (i.e., the message format standards) where there is uniform movement of healthcare data from one system to another such that the clinical or operational purpose and meaning of the data is preserved and unaltered. Structural interoperability defines the syntax of the data exchange. It ensures that data exchanges between information technology systems can be interpreted at the data field level.
- "*Semantic*" interoperability provides interoperability at the highest level, which is the ability of two or more systems or elements to exchange information and to use the information that has been exchanged (IEEE 1990). Semantic interoperability takes advantage of both the structuring of the data exchange and the codification of the data including vocabulary so that the receiving information technology systems can interpret the data. This level of interoperability supports the electronic exchange of patient summary information among caregivers and other authorized parties via potentially disparate electronic health record (EHR) systems and other systems to improve quality, safety, efficiency, and efficacy of healthcare delivery (HIMSS 2010).

## 14.11  Telemedicine

Telemedicine has continued its steady rise in 2016. Ralph D. Derrickson, CEO of Carena, a virtual care provider saw that growth accelerating in 2017. He predicted that "Factors like the rise in learner, more expensive health plans, value-based demands placed on providers, the new Medicare Access and Chip Reauthorization Act (MACRA) and 'consumerization' will all drive more rapid adoption in 2017." More advanced technology, plus IoT will provide and enable "ever-increasing" ways for providers to remotely diagnose as well as treat patients. For example, doctors can remotely monitor heart rate, and blood pressure using wearables. The data collected can aid in diagnostics, disease management, and facilitate treatment of chronic illnesses (https://www.health-datamanagement.com/list/top-healthcare-it-trends#slide-11).

Smith (2015) tried to distinguish telemedicine and telehealth in this brief titled "Telemedicine VS Telehealth: What's the Difference?" People in the industry use both terms interchangeably, but Smith draws the distinction below to separate them.

First, telehealth is a subset of E-Health. E-Health uses the internet and telecommunications to provide a vast array of services, such as the "delivery of health information, for health professionals and health consumers, education and training of health workers and health systems management."

Telehealth, she opined, includes "a broad range of technologies and services to provide patient care and improve the healthcare delivery system as a whole." Telehealth, thus, refers to both clinical services and nonclinical services such as remote non-clinical services, including provider training, administrative meetings, and continuing medical education, in addition to clinical services. The World Health Organization (WHO), indicates that telehealth also, includes, "surveillance, health promotion and public health functions." Telehealth, is more encompassing and refers to a broader scope of remote healthcare services than does telemedicine, which refers specifically to remote clinical services.

Telemedicine, then, is simply a subset of telehealth that refers solely refers to the use of telecommunications technology to provide healthcare services and education over a distance. Additionally, telemedicine involves the use of electronic communications and software to provide clinical services to patients without an in-person visit. Telemedicine technology can often be used for follow-up visits, management of chronic conditions, medication management, or specialist consultation among other clinical services that can be provided remotely through audio and video connections that are secured.

Finally, the WHO also uses the term *telematics* which for health is a composite term for both telemedicine and telehealth. It can also be used for any health-related activities carried out over distance using information communication technologies.

In conclusion, all telemedicine is telehealth, not all telehealth is telemedicine. They constitute part of the larger effort, Smith (2015) concludes "to expand access to care, make health management easier for patients and improve the efficiency of the healthcare delivery network."

### 14.11.1   Telemedicine and Telehealth: HRSA

Here is another definition of telehealth provided by the Health Resources and Services Administration (HRSA) of the U.S. Department of Health and Human Services. Telehealth is "the use of electronic information and telecommunications technologies to support and promote long-distance clinical healthcare, patient and professional health-related education, public health and health administration. Technologies include videoconferencing, the internet, store-and-forward imaging, streaming media, and terrestrial and wireless communications." The definition of telemedicine is subsumed and they do not bother to differentiate them.

*Among the Telehealth applications are* (HealthIT 2017):

- *Live (synchronous) videoconferencing*: a two-way audiovisual link between a patient and a care provider
- *Store-and-forward (asynchronous) videoconferencing:* transmission of a recorded health history to a health practitioner, usually a specialist.
- *Remote patient monitoring (RPM):* the use of connected electronic tools to record personal health and medical data in one location for review by a provider in another location, usually at a different time.
- *Mobile health (mHealth):* healthcare and public health information provided through mobile devices. The information may include general educational information, targeted texts, and notifications about disease outbreaks.

## 14.12   Conclusion: Technology/Decisions

Portable health records also imply that the Providers of services can share and exchange information on a patient's behalf. Because the health record will reside in the Cloud or some secured place, only those authorized can access and download on their mobile devises, thereby offsetting security and privacy issues. The commonality of language (semantics and syntactics) will become crucial.

How will clinical and business decision support be enabled using Portable Health Records? Clinical decision support is helping care providers make decisions about the care for individual patients based on an available clinical knowledge base. Business decision support is helping administrators make informed decisions based on the analysis of the organizations specific mix of patients and a business knowledge base.

### 14.12.1   Health Technology Decisions

A deluge of technologies that are being applied to healthcare and those with potential applicability in the future of healthcare have been described. While they present several opportunities, they could also lead to decision paralysis for those who must make those choices.

Nevertheless, artificial intelligence seems like a viable option—but only to the extent that it is affordable! Blockchain holds some promise, at least, it alleviates the problems of interoperability and holds the promise for security and privacy of data….except that it is still not a mature technology. It needs to address challenges that are outside healthcare, such as technical, organizational, and behavior challenges.

Cloud computing is moving at an accelerated speed. But, the applications so defined, don't quite meet the attributes required of Cloud computing. They are moreover burdened by privacy and security issues.

The Internet of Things (IoT) seems so close to resolving the issues because it is virtually "a connected set of anyone, anything, anytime, anyplace, any service, and any network." It would therefore make time and place irrelevant for both provider and patient in the healthcare field. But then, what does it operate on—Cloud computing? Cloud computing when in association with the Internet of Things (IoT), nothing but obfuscation results.

Telemedicine and imaging technologies have limited application areas and are considerably pricy. So are disease management technologies and electronic health records.

So, the best advice to a healthcare provider who aspires to invest in technology for the delivery of healthcare would be: "Buyer Be Ware". There is so much fuzziness in the field, many competing and complicating demands, and no strategic direction towards elimination of waste, reduction of costs, or more affordable healthcare—at least, not yet in the US!

### 14.12.2 Conclusion and the Way Forward

Today's EMR/EHR is no closer to *portability*, than yesterday's was. It operates in the U.S. healthcare environment that is riddled with inefficiencies and high cost of delivery. It has no defined route/s for access (in all its dimensions). It is hindered by various laws, legal underpinnings, business interests and patient's acceptance.

In the introductory section, it was noteworthy to examine the forms and mechanisms that are currently used and that will be used to provide portability and universal access to persons' medical or health records. The discussion that ensued covered the form, access, the law and some type of report card for the U.S. in comparison to ten other developed countries. The U.S. report card on healthcare was abysmal. What the U.S. has set up as its healthcare system is unsustainable because of so many factors. Amongst them are fragmentations, the outcome of which are chronic inefficiencies and escalating costs that result in poor quality of care. It is therefore, time to regroup. The evolutionary steps, all be it, incremental changes and purported improvements are doomed to fail. What is needed in the U.S. healthcare is a *Revolution!*

## References

Bean R, Stephen G. How Blockchain is impacting healthcare and life sciences today. Jersey City: Forbes; 2018.

Bresnick J. HealthIT Analytics: top 12 ways artificial intelligence will impact healthcare. 2018. https://healthitanalytics.com/news/top-12-ways-artificial-intelligence-will-impact-healthcare. Accessed 24 Jul 2018.

Ehteshami BB, Veta M, van Diest JP, van Ginneken B, Karssemeijer N, Litjens G. Diagnostic assessment of deep learning algorithms for detection of lymph node metastases in women with breast cancer. JAMA. 2017;318(22):2199–210. https://doi.org/10.1001/jama.2017.14585.

Griebel L, Prokosch HU, Köpcke F, Toddenroth D, Christoph J, Leb I, Engel I, Sedlmayr M. A scoping review of cloud computing in healthcare. BMC Med Inform Decis Mak. 2015;15(1):17. https://doi.org/10.1186/s12911-015-0145-7.

Health Information Technology. 9 things to know about blockchain in healthcare (Becker's Health IT & CIO Report). 2016. https://www.beckershospitalreview.com/healthcare-information-technology/9-things-to-know-about-blockchain-in-healthcare.html.

HealthData. 10 top healthcare information: artificial intelligence. 2016a. https://www.healthdata-management.com/list/top-healthcare-it-trends#slide-2. Accessed 24 Jul 2018.

HealthData. 10 top healthcare information: new trends accelerate change within healthcare IT. 2016b. https://www.healthdatamanagement.com/list/top-healthcare-it-trends#slide-1. Accessed 24 Jul 2018.

HealthIT. Telemedicine and telehealth. 2017. https://www.healthit.gov/topic/health-it-initiatives/telemedicine-and-telehealth.

HealthIT. Medical practice efficiencies & cost savings. 2018. https://www.healthit.gov/topic/health-it-basics/medical-practice-efficiencies-cost-savings.

HIMSS. HIMSS dictionary of healthcare information technology terms, acronyms and organizations. Appendix B. 2nd ed. Chicago: HIMSS; 2010. p. 190.

HIMSS. HIMSS dictionary of healthcare information technology terms, acronyms and organizations. 3rd ed. Chicago: HIMSS; 2013. p. 75.

HIMSS17 Infocus. The interoperability imperative: value-based care depends on health information exchange. 2017. http://www.healthcareitnews.com/himss-infocus/interoperability?utm_campaign=himss-infocus&utm_medium=text_ad&utm_source=himssorg&utm_term=infocus-interop.

Institute of Electrical and Electronics Engineers. IEEE standard computer dictionary: a compilation of IEEE standard computer glossaries. New York: IEEE; 1990.

Islam SR, Kwak D, Kabir MH, Hossain M, Kwak KS. The internet of things for healthcare: a comprehensive survey. IEEE Access. 2015;3:678–708. https://www.researchgate.net/publication/280696619_The_Internet_of_Things_for_Health_Care_A_Comprehensive_Survey. Accessed 06 July 2018.

Jones SS, Adams JL, Schneider EC, Ringel JS, McGlynn EA. Electronic health record adoption and quality improvement in US hospitals. Am J Manag Care. 2010;16(12 Suppl HIT):SP64–71.

National Committee on Vital and Health Statistics (NCVHS). Report on uniform data standards for patient medical record information. Hyattsville: NCVHS; 2000. p. 21–2.

Norman A. The stethoscope of the 21st century. 2018. https://futurism.com/ai-medicine-doctor/. Accessed 24 Jul 2018.

Shashank A. 6 reasons why healthcare needs the internet of things (IoT). 2017. file:///Users/Mariposa/Desktop/Editor-Co-Ed%20Springer%20Reviewed%20papers%20from%20March18%20/Dr%20Egondu%20Onyejekwe%20-%20Springer/Completed%20Drafts%20/New%20References%20to%20Tech%20and%20Decisions/6%20Reasons%20Why%20Healthcare%20Needs%20The%20Internet%20of%20Things%20(IoT).htm.

Smith A. Telemedicine vs. telehealth: what's the difference? Austin: Chiron Health; 2015. https://chironhealth.com/blog/telemedicine-vs-telehealth-whats-the-difference/.

The Daily Beast. Paging Dr. AI: why doctors aren't afraid of better, more efficient AI diagnosing cancer. Just like humans, AI isn't perfect. 2017. https://www.thedailybeast.com/why-doctors-arent-afraid-of-better-more-efficient-ai-diagnosing-cancer.

Thota C, Sundarasekar R, Manogaran G, Varatharajan R, Priyan MK. Centralized fog computing security platform for IoT and cloud in healthcare system. In: AVK P, editor. Exploring the convergence of big data and the internet of things. Hershey: IGI Global; 2018. p. 14. https://www.igi-global.com/chapter/centralized-fog-computing-security-platform-for-iot-and-Cloud-in-healthcare-system/187898.

Zhang P, Walker MA, White J. Metrics for assessing blockchain-based healthcare decentralized apps. In: IEEE 19th international conference on e-health networking, applications and services (Healthcom): IEEE; 2017.

# Chapter 15
# Security

**Hung Ching**

**Abstract** Smart technology is not necessarily invulnerable and impregnable. The threat is real, and cyber crime is one of the most important issues in the realm of portable healthcare. Security and privacy are entwined when dealing with portable health records. Solutions against cyber crime are constantly being created to combat the would-be thieves. Electronic signatures, elliptic curve cryptography, and blockchain-based storage methods are all ways that healthcare professional can use to prevent theft of information in this electronic age.

**Keywords** Security · Threats · Cyberthreats · Cyber criminals · EHRs · Electronic signature · Cryptography · Blockchain

## 15.1 Introduction

Various electronic devices in our mobile society such as cell phones and laptops come supplied with software that enables them to be smart. However smart technology is not necessarily invulnerable and impregnable. Any smart apparatus that is connected to the internet can potentially be hacked, and all of the information and content of the portable smart device can be stolen and compromised. The threat is real, and cyber crime is one of the most important issues in the realm of portable healthcare.

H. Ching (✉)
Department of Medical Physics, Memorial Sloan Kettering Cancer Center (National Institutes of Health Funded), New York, NY, USA
e-mail: chingh@mskcc.org

© Springer Nature Switzerland AG 2019                                                      189
E. R. Onyejekwe et al. (eds.), *Portable Health Records in a Mobile Society*,
Health Informatics, https://doi.org/10.1007/978-3-030-19937-1_15

## 15.2 Hardware and Software Threats

Security and privacy are entwined when dealing with portable health records. Threats are constantly at play. Examples of hardware threats are theft of portable smart devices and theft of computers that had been connected to smart devices via the internet or via universal serial bus (USB) cable. Examples of software threats are theft of portable electronic health records (EHRs) from smart devices when they are connected to the internet and theft of EHRs from computers which tend to be always connected to the internet. EHRs can also be stolen via direct downloading of information from these computers using a flash drive. Threats can be external as exemplified by an outside hacker or a foreign entity. Threats can also be internal as exemplified by an employee who steals EHRs or who simply is just careless or does not follow protocol. All of these examples may lead to security and privacy breaches that may have dire consequences.

## 15.3 Internal Threats

Threats to the security and privacy of portable EHRs often originate from internal sources. A person can be convinced via financial incentive or blackmail to steal information by accessing information on the computer networks located at his or her workplace. The workplace can be a health care facility such as a hospital, clinic, or doctor's office. The theft of EHRs can easily be done by simply downloading EHRs to a flash drive and passing it on to the people that will pay handsomely for it.

Internal threats do not have to involve corruption or selling of stolen EHRs. An employee who works at the health care facility can be a very honest person, but what if he or she neglects to follow company protocol on cyber security? By being careless, the employee may have just become the weakest link at the workplace. The people and organizations who represent the external threats seek to break in or channel in via the weakest link to steal EHRs.

## 15.4 External Threats

In addition to internal threats, threats to the security and privacy of portable EHRs can also originate from external sources. Hackers raid portable devices to demonstrate their ability to infiltrate these devices. Often the motivation for hackers is simply the recognition they will receive among their peers (Connexion Healthcare 2013). It is now easier than ever to become a hacker because there are attack protocols available on the internet for anybody to download. If the motivation for hackers is simply fame and prestige, imagine how much more motivated they would be if they were offered large financial incentives.

Monetary gains catapult us into the realm of cyber crimes. They are often perpetrated by criminals who raid portable devices in order to steal portable EHRs. Cyber criminals infiltrate portable devices via the use of spyware when they successfully trick potential victims to download contents of the spam that they send out (Connexion Healthcare 2013). With the stolen portable EHRs, cyber criminals can commit identity theft and fraud. There is a special kind of malware called ransomware that cyber criminals utilize to commit extortion. The ransomware prevents access to the portable device until a ransom is paid by the targeted individual or health organization.

Who better to use spyware than an actual spy? Employed by foreign governments, agents raid portable devices to benefit their motherland. The motivation here is mostly political and economic (Connexion Healthcare 2013). With stolen information from the individual or health organization, agents of foreign government can cause great disruption and inconvenience in our mobile society. Foreign governments may even sell the stolen information to other foreign governments or enterprises.

Finally, terrorists should not be excluded from this list of external threats. As with everything else that terrorists do, they have a tendency to destroy what is good. A mobile society using portable health records would be subject to attacks by terrorists (Connexion Healthcare 2013). The goal of terrorists is to totally annihilate the workings of a mobile society to create fear among people and to negate the assumption that people live in a safe environment.

## 15.5   FDA on the Record in 2013

The problem with cyber security for medical devices and hospital networks has become so important that the Food and Drug Administration (FDA) issued a report on June 14, 2013 recommending specific safeguards to lower the risk of loss and breakdown due to cyberattack. The FDA specifically mentioned the issues of unauthorized installation of malware in medical devices and unwarranted access to medical equipment and hospital computer and wireless networks as the main culprits (FDA 2013). The FDA identified "the presence of malware on hospital computers, smartphones and tablets, targeting mobile devices using wireless technology to access patient data, monitoring systems, and implanted patient devices," (FDA 2013) as one of the key cyber security breaches. Hospital smartphones and tablets were noted by the FDA as possible devices where patient data can be stolen from. Just imagine how enormous the issue can be with the inclusion of all patient smartphones and tablets which can be targeted by hackers, cyber criminals, foreign governments, and terrorists!

## 15.6   FDA on the Record in 2016

In a joint collaboration with the Department of Homeland Security (DHS), the National Health Information Sharing Analysis Center (NH-ISAC), and the Department of Health and Human Services (DHHS) which the FDA is a part of, the FDA stated, "while the increased use of wireless technology and software in medical devices also increases the risks of potential cyber security threats, these same features also improve health care and increase the ability of health care providers to treat patients" (FDA 2016). From this very far-reaching statement by the FDA, it is understood that all stakeholders are to help in the effort to find the balance between ensuring patient safety and developing improved technologies in our mobile society. One of the key messages from this collaboration is the FDA's recommendation that health care facilities such as hospital, clinics, and doctor's offices, should constantly assess their network security in order to protect patients and their portable EHRs.

## 15.7   Minimizing Threats

Even though portable smart devices are assuming the role of personal computers in this day in age, security solutions for portable smart devices are not as developed and are not as popular as those security solutions for personal computers. Patients who own portable smart devices such as cell phones and tablets can take precautionary measures to ensure that their information remains private and protected. Following the best practices concerning security for portable smart devices may not completely eliminate cyber threat, but it can surely lower the risk of a potential cyber attack. Below are specific security measures that can be taken by the patient to ensure cyber security on their portable smart devices.

One of the most important examples of best practices for cyber security of portable smart devices is to act quickly and report loss or theft of these devices immediately. Twenty-four hour toll-free hotlines for all mobile service providers in the United States are available to immediately report such events. Mobile service providers can limit access and use of the device by denying the device connectivity to the internet and cellular service. In this way, ill-intentioned misuse of the device can be effectively prevented.

Stopping internet connectivity and cellular service may stop unauthorized access and use of what is on the hard drive of the portable smart device up to a certain extent. In order to prevent access to data that is on the hard drive of portable smart devices, some mobile service providers have features on devices that allow them or the owner of the devices to remotely delete all of the data from the devices. This feature is called remote wiping (US-CERT 2016).

Similarly, remote wiping is very useful when a malicious app has been installed on a portable smart device. These kinds of apps have a tendency to take control of the device and render it useless and inaccessible to the owner of the device. The owner may be unaware that the device has been hacked and may delay visiting or calling the service provider on the 24 hotline. It such instances, it is always advisable to seek the help of the service provider immediately. Having the experts perform a remote wipe to rid the device of a malicious app can potentially save further pain and aggravation.

Another feature on smart portable devices is password protection. It is a feature that most of us assume is a must have on desktops in the office environment. However, password protection is still a feature that is not fully utilized by everybody. Just imagine how easy a hacker and not to mention any arbitrary person can obtain access to the portable device and the information that is contained within it.

Just as we tend to be suspicious of unrecognizable emails and unfamiliar websites that can potentially ruin desktops and computers, we must be equally vigilant about them when it comes portable smart devices. Limiting exposure of the portable device to cyber criminals is key to preventing the smart device from being hacked. Unfamiliar free Wi-Fi is also something that we must be cautious of. They can be purposely set up to steal information from portable smart devices. Finally, it is good practice to stay away from public Wi-Fi as cyber criminals are on the constant prowl for potential victims.

## 15.8 Failed Attempt at Portable Health

Back in 2008, Google launched Google Health, which was the company's solution to portable EHRs. Google Health was a failure, and in 2013 Google systematically destroyed all of the data that had been left on all user accounts (Google 2013). What were the main issues that were responsible for the failure? The first issue is that Google Health did not have fun or engaging (Mobilehealthnews 2011) content that other social media apps have. It was not enough to just strive on having users staying healthy since users were looking for something more. What Google Health did not succeed in doing was to generate peer pressure among the population to make everyone want to be a part of it.

The second issue for Google Health was that users did not find it to be trustworthy. Users thought that Google Health should not be trusted with so much personal data especially private health information. Some users even thought that Google had some ulterior motives for collecting the health data (Informationweek 2011). The third issue involved complaints that Google Health was too cumbersome and sometimes too difficult to maintain (Mobilehealthnews 2011). To compound the problems, users were receiving information that was sometimes not reliable, so they simply lost interest in updating information on Google Health.

Google Health's failure can be traced to a fourth issue. Google Health did not have health practitioners on board. Even if it had a massive following, Google Health would have lost it without the leadership of clinicians (Mobilehealthnews 2011). A fifth issue that had arisen dealt with health insurance companies. Google was frustrated in trying to convince health insurance companies to voluntarily share their health information data (Mobilehealthnews 2011). Finally, the last issue was that Google Health lacked convenience features such as the ability to make a doctor's appointment for the user or to even get a prescription refilled (Informationweek 2011).

There are currently other solutions to portable EHRs such as Microsoft's HealthVault. Companies such as Microsoft need to learn from their own experiences and those of Google Health on how they can make a better and more user-friendly product. This must be done in the interest of self-preservation so that these products do not eventually share the same fate as that of Google Health.

## 15.9   Double Trouble Caused by EHRs

EHR lawsuits have increased over the years as exemplified by the statistics collected by a national malpractice insurance company called the Doctors Company. From the company data, EHR-related lawsuits consisted of only 1% of lawsuits from 2007 to 2013. In between 2013 and 2014, EHR-related lawsuits doubled (Medical Economics 2015). It is anticipated that this figure will go even higher in subsequent years as more and more clinicians are using EHRs. Errors in data input from the physician's office and other user errors were the primary reasons for EHR-related lawsuits. In the period of time from 2007 to 2014, 64% of the EHR-related lawsuits from the Doctors Company were attributed to user errors. In that same period time, 42% of the EHR-related lawsuits involved system factors (Medical Economics 2015).

## 15.10   Challenges and Obstacles

Clinicians typically are not allowed to bill for services except for face-to-face visits (Commonwealth Fund 2011). This means that clinicians cannot bill for services they render through information that is sent to them through mobile apps. Unfortunately, health insurance companies are in no hurry to change this fee-for-service business model anytime soon. In order for mobile health technology to continue to be used by clinicians to help patients, it is imperative that clinicians be reimbursed for services rendered through mobile apps as if they were face-to-face visits.

   As with any new technology, there is always people who get left behind because of their refusal to use innovative technology or because of their inability to afford devices. In the case of mobile health, not being able to afford smart phones or tablets is one of the issues that patients are dealing with. On top of that, there is an additional issue of not being tech savvy enough to use the mobile apps (Commonwealth Fund 2011). The question then is how is it possible to bridge the digital divide? One such solution is to utilize text messaging platforms from either a smart phone or a regular mobile phone to convey the necessary information between clinicians and patients. Just as mobile in mobile phone means 24/7 access for communication between people without borders, mobile in mobile health could potentially mean 24/7 access to healthcare for patients with boundaries.

## 15.11  Electronic Signatures

Electronic signatures are not just mere digital renditions of handwritten signatures. The electronic signature carries extra dimensions when compared to its handmade counterpart. A digital signature can be treated as an electronic signature only when it conforms to industry standards and legal regulations. The electronic signature can be in any form such as an electronic sound or electronic symbol. Electronic signature software grants users the ability to share health data in a legitimate and secure method with patients and third parties. The software used has to be compliant with the Health Insurance Portability and Accountability Act (HIPAA), the Patient Protection and Affordable Care Act (PPACA), and requirements set forth by the Joint Commission on Accreditation of Health Care Organizations (JCAHO) (McTosh 2014).

Electronic signatures can help in streamlining processes such as the process of obtaining patient consent. If patients are not physically present at a medical facility, they typically have print and sign a patient consent form, and then fax the consent form back to the facility. This can be indeed an inconvenience. Electronic signatures can facilitate the entire patient consent process by allowing patients to use their portable devices such as their smart phones or tablets to complete the patient consent process.

By utilizing electronic signatures, healthcare facilities can lower the the risk of loss due to damaged or improperly filled out forms. Electronic signatures can help to reduce delays in the process of filling out forms. Software used in conjunction with electronic signatures can guide the patient or healthcare professional through the entire process making sure that all "paperwork" adheres to protocol. After all documents are correctly filled out, the software can seal the documents with a tamper-evident electronic seal (McTosh 2014).

## 15.12  Elliptic Curve Crytography

Elliptic curve cryptography (ECC) is a group of algorithms for encryption and de-encryption of information and for the exchange of cryptographic keys (Sullivan 2013). ECC has found much success in protecting EHRs while it is being electronically delivered to a trusted destination. Moreover, ECC has even been applied in protecting real-time monitoring and transmission of physiological signals and medical images (Tsai et al. 2014).

## 15.13  Blockchain-Based Storage Methods

An innovative method to secure EHRs against cyber attacks is the utilization of a blockchain-based storage system for the sharing and maintenance of clinical information. Blockchain had been, at the outset, conceived as an infrastructural part of Bitcoin, the cryptocurrency that has made headlines in recent times (Ivan 2016).

Blockchain-based technology is based on three fundamental concepts. The first tenet is that information is stored in a public ledger that everyone can read. The contents of the ledger can never be altered nor deleted. The second tenet is that the blockchains are set up in a decentralized chain of computerized nodes, which make them very resistant to cyber attacks. No company or person controls a blockchain since it is public and decentralized. The third tenet is that the data associated with each transaction on blockchain is accessible to everybody that is on the system, but this does not necessarily imply that the information is readable as the data is heavily encrypted. A process of changing names to only identifiers, called pseudoanonymity, and public key infrastructure (PKI) are both used to encrypt data in blockchain, making it virtually impossible and very expensive to decipher (Ivan 2016).

## 15.14   Conclusion

In conclusion, cyber threat is one of the most important problems in the realm of portable healthcare. There are various creative and non-creative ways that cyber criminals can steal information from electronic devices such as a desktop computer, laptop, tablet, and smart phone. Solutions against cyber crime are constantly being created to combat the would-be thieves. Electronic signatures, elliptic curve cryptography, and blockchain-based storage methods are all ways that healthcare professional can use to prevent theft of information in this electronic age.

## References

Commonwealth Fund. Health care quality improvement: there's an app for that. 2011. http://www.commonwealthfund.org/publications/newsletters/quality-matters/2011/october-november-2011/in-focus. Accessed 14 May 2018.

Connexion Healthcare. Cyber security and mobile medical devices: protecting and securing patient medical information. 2013. http://connexionhealthcare.com/wp-content/uploads/2013/04/Cyber-Security-and-Mobile-Medical-Devices.pdf. Accessed 14 May 2018.

Food and Drug Administration (FDA). Cybersecurity for medical devices and hospital networks: FDA safety communication. 2013. http://www.fda.gov/MedicalDevices/Safety/AlertsandNotices/ucm356423.htm. Accessed 14 May 2018.

Food and Drug Administration (FDA). Cybersecurity. 2016. http://www.fda.gov/MedicalDevices/DigitalHealth/ucm373213.htm. Accessed 14 May 2018.

Google. Google health has been discontinued. 2013. https://www.google.com/intl/en_us/health/about/. Accessed 14 May 2018.

Informationweek. 5 reasons why Google Health failed. 2011. http://www.informationweek.com/healthcare/electronic-health-records/5-reasons-why-google-health-failed/d/d-id/1098623?. Accessed 14 May 2018.

Ivan D. Moving toward a blockchain-based method for the secure storage of patient records. 2016. https://www.healthit.gov/sites/default/files/9-16-drew_ivan_20160804_blockchain_for_healthcare_final.pdf. Accessed 14 May 2018.

McTosh P. Implementing electronic signatures. 2014. https://www.hitechanswers.net/implement-e-signatures-ehrs/. Accessed 14 May 2018.

Medical Economics. Avoiding an EHR-related malpractice suit. 2015. http://medicaleconomics.modernmedicine.com/medical-economics/news/avoiding-ehr-related-malpractice-suit. Accessed 14 May 2018.

Mobilehealthnews. 10 reasons why Google Health failed. 2011. http://mobihealthnews.com/11480/10-reasons-why-google-health-failed. Accessed 14 May 2018.

Sullivan N. A (relatively easy to understand) primer on elliptic curve cryptography. 2013. https://arstechnica.com/information-technology/2013/10/a-relatively-easy-to-understand-primer-on-elliptic-curve-cryptography/. Accessed 14 May 2018.

Tsai KL, Leu FY, Wu TH, Chiou SS, Liu YW, Liu HY. A secure ECC-based electronic record system. J Internet Serv Inform Sec. 2014;4(1):47–57.

United States Computer Emergency Readiness Team (US-CERT). Cyber threats to mobile phones. 2016. https://www.us-cert.gov/sites/default/files/publications/cyber_threats-to_mobile_phones.pdf. Accessed 14 May 2018.

# Chapter 16
# FITT Model

**Dasantila Sherifi**

**Abstract** As more healthcare technologies become available and accessible through mobile devices, it becomes important to evaluate them and explore their potential for adaption in various environments or identify adaption problems. One of the technology adaption theories is the Fit between Individual, Task, and Technology (FITT) by Ammenwerth, Iller, & Mahler. This theory takes into account the attributes of individuals, attributes of tasks and processes, and attributes of technology. This chapter will explain the FITT framework and the application of FITT theory in evaluating clinical healthcare systems. Better understanding of the FITT theory, especially individual-task, task-technology, and individual-technology interactions and fit among them, along with better understanding of their impact on successful adaptation of new technologies may be helpful for those engaged in designing, analyzing, or implementing mobile technologies in healthcare.

**Keywords** FITT model · FITT framework · Technology adaption · Information technology adaption · Technology fit model · FITT application

## 16.1   Introduction

Health information technologies have grown tremendously and according to Markets and Markets (2015), by 2020, the growth in the healthcare IT industry in North America is predicted to reach $104.3 billion. Healthcare information technologies, including mobile ones are being used by providers and patients at different levels. Research shows various rates of adaption and usage of information technologies, as well as successful, marginal or failed implementations of such technologies. While there are a number of theories that explain variation in adaption and usage, this chapter presents a theory that explains adaption of technologies in healthcare from a

D. Sherifi (✉)
College of Health Sciences, DeVry University, Naperville, IL, USA
e-mail: dsherifi@devry.edu

© Springer Nature Switzerland AG 2019                                        199
E. R. Onyejekwe et al. (eds.), *Portable Health Records in a Mobile Society*,
Health Informatics, https://doi.org/10.1007/978-3-030-19937-1_16

perspective of Fit between Individual, Task, and Technology. Most importantly, the chapter demonstrates how the FITT Model could be used in the process of healthcare IT implementation in order to make the implementation more successful.

## 16.2　Background

The case for using healthcare technologies has already been made and patients' and providers' efforts in adapting those technologies reflect that. A 2015 national survey showed that 58% of smartphone users had downloaded and/or used mobile health apps (Krebs and Duncan 2015). In addition, the number of physicians using medical apps grew from 50% in 2010 to 70% in 2015 (Krebs and Duncan 2015). Despite the promising growth of mobile health apps in general, use of certain applications is still low. According to MedDataGroup (2016), while 60% of physicians use mobile access to EHR, only about 15% use patient evisits, less than 25% use mobile devices for health monitoring, and about 35% use patient portals for scheduling communication. The importance of health information technologies is not only recognized by providers and the general public but also by the government. In the 2011–2015 federal health information technology strategic plan, the Office of National Coordinator for Health Information Technology (ONC) included a specific goal to achieve rapid learning and technological advancement (2017). The vision is that new technologies will be used to collect and analyze data quickly and efficiently, so that the knowledge acquired from them can be used immediately. In order for such expectations to be achieved, information technologies need to be adapted and used optimally (not just marginally). The reality is that many providers use a certain function of the system but do not use another (because they do not know exactly how to use or perhaps it is cumbersome). Many print a report from one system, and then, scan it into another (which seems to defeat the purpose of the health IT). Many patients download a healthcare app on their phone but do not enter health data in it. Government regulations and requirements can impact implementation of healthcare technologies but they may not be sufficient to sustainably foster their adaption in a meaningful and efficient way among users.

## 16.3　Explaining Adaption of New Information Technologies

Since the emergence of new technologies, use of computers, or computer applications, researchers have come up with theories and models that explain adaption of such technologies and applications. In 1962, Everett Rogers came up with the Diffusion of Innovations theory which rests on the premise that an innovation is successfully adapted when it has a perceived relative advantage, is compatible with existing values and practices, is perceived as simple and easy to use, can be tried without risk, and provides observable results (Rogers 2003). In 1989, Davis, Bagozzi and Warshaw developed the Technology Acceptance Model (TAM) which is derived from the Theory of Reasoned Action and explains voluntary usage of an information system or computer technology as a function of perceived usefulness and perceived ease of use for the particular computer technology (Davis et al. 1989). Perceived usefulness and perceived ease of use lead to a certain attitude toward using the system, which in turn leads to behavioral intention to use the system, and ultimately affects the actual use or adoption of the system. Growth of information systems and the requirements to use various applications in the workplace spurred interest in exploring the fit between the task and technology, which gave birth to the Task-Technology Fit (TTF) Model by Goodhue in 1995. This model considered the characteristics of the task to be accomplished, the technology available to complete it, and the fit between the task and the technology, which in turn could affect utilization or performance of the technology (Goodhue and Thompson 1995). Ammenwerth, Iller, and Mahler took these theories a step further by adding the individual component and developed the Fit between Individual, Task and Technology.

## 16.4   FITT

The Fit between Individual, Task and Technology (FITT) Model was developed in 2006. The intention of FITT was to provide a framework that would help analyze the adoption process of health information technologies, mostly in clinical settings. Ammenwerth et al. (2006) estimated that about 60–70% of software implementation projects failed and that perception regarding success and failure varied among different healthcare settings, departments, and roles. According to the FITT framework, adoption of information technologies in healthcare depends on three main factors: attributes of the individual, attributes of the clinical tasks and processes, and attributes of the technology (Ammenwerth et al. 2006). The framework goes beyond explaining the socio-organizational factors pertaining to IT adoption; it provides a platform to explore the user, the task, and the information system at hand and see if they are in synch. The level of fit predicts the success of IT adoption.

Attributes of the individual represent characteristics of a single user or a user group (Ammenwerth et al. 2006). They include various aspects from computer anxiety and motivation to knowledge about the information system, interest of the user(s) on the task at hand, flexibility, collaboration, and communication pertaining to team work, organizational context, or even politics in the workplace. Attributes of the clinical tasks and processes represent elements such as organizational aspects, work flow, or task complexity. It's important to note that the task component involves all activities about the particular task, as well as their interdependence. Attributes of the technology represents usability, functionality, and performance of the technology. The key with the technology component is that it does not simply refer to the information system but the computer, network infrastructure, integration of applications, and other tools (even paper that may be involved) which are supported from that particular technology, as well as their reliability. Figure 16.1 visualizes how the three components of the FITT Model connect to each other and collectively contribute to the IT adoption.

**Fig. 16.1** The FITT framework (By Ammenwerth et al. 2006)

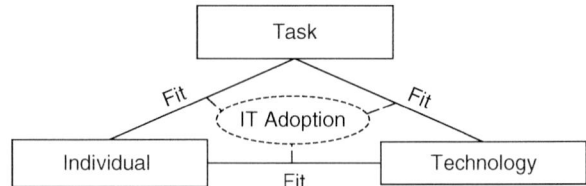

According to the model, fit can be influenced and improved by directly influencing the attributes of each of the three components. Interventions can be undertaken at the individual, task, or technology level (Ammenwerth et al. 2006). Following are some examples.

In order to intervene at the individual level, users can be involved in the process of application selection through participation in demos, providing feedback, or piloting the product. Training about computer usage in general and effective training on the new applications can improve their knowledge and understanding of the system. Effective communication can also play an important role when it comes to attitudes toward the system or sharing updates or issues related to it. Interventions at the task level can require reorganization of the task(s), changes in the workflow, as well as changes or clarifications in the responsibilities involved, such as who is to record the data or document certain details. Technology level interventions may require changes in the design of the system work flow, screen design, hardware and network updates, or integration of certain parts of the system or among systems.

FITT framework is an applied model, which means it is already tested and evaluated (Logan 2012). In 2011, Honekamp and Ostermann used FITT to evaluate health information systems prototypes based on results from patient-system interactions (Honekamp and Ostermann 2011). Their quantitative study revealed that the evaluation using FITT is suitable for evaluating new health information systems. Lesselroth used FITT to develop a survey that measured three variables: provider attitudes toward the task at hand—medication reconciliation, provider's perceptions of health information technology, and the local organizational climate for implementation (Lesselroth et al. 2011). The data collected enabled evaluation of provider perceptions on the new information system and revealed certain associations between provider attitude, provider perceptions, and implementation climate, as well as implementation effectiveness.

## 16.5   FITT and SDLC

FITT can be used at various stages of the systems development life cycle, which typically consist of planning, analysis, design, implementation and evaluation (Sayles 2013). The use of FITT in these stages is illustrated through the following examples and explanations.

The planning stage involves activities such as discussion of the system/technology needed, required, or wanted, as well its costs, the actual budget, and timing for implementation (Sayles 2013). This is considered the beginning of the project. Implementation of FITT in this stage could start with a broad analysis of the tasks that need to be performed via the new application and how well the application would be expected to perform such tasks.

The analysis stage requires a thorough investigation of the details on what exactly the system will do, what will be impacted, or what else is needed (Sayles 2013). Most importantly, the use case and sequence diagrams are created. The use case diagram visualizes who will use the system and how they will use it. For example, usage expectations for a patient would be to simply download a prescription refill mobile application and scan/type the prescription number. On the other side, the expectation of usage for the pharmacist would be to not only receive the information but to also check for prescription or allergy updates in the system, and then, refill the prescription. The sequence diagram visualizes the steps needed to complete the task. By detailing who will use the system and how they will use it, it is possible to understand the required level of user knowledge and skills, and the details of the task. Disruptions at the point of planning and analysis can be described and analyzed with regard to disruptions in any of the three fit dimensions (task-technology, technology-individual, or individual-task). Consideration of these "fits" improves planning and creates an opportunity to anticipate and prepare in advance.

Activities that occur in the design phase are mostly related to specifying the system functions and creating screen prototypes. Interventions or changes in the system, screen design, integration, data transmission, etc. will affect the fit dimensions, and therefore, can be evaluated from a perspective of the fit with users and fit with task from the beginning. Interventions in this phase may also depend on whether the organization is developing the system in house via a Joint Application Development (JAD) process or purchasing it from vendor. The first option allows more flexibility for adjusting the individual-technology and task-technology fits. For example, the current design requires pushing "Tab" after entering the data values for each data element; yet, users are used to pushing the "Enter" button. Identifying this particular lack of fit allows for an adjustment in design and better fit between the individual and technology. On the other side of the triangle, consider the task of registering the patient and the design that is available to accomplish this task. A patient can be identified and registered without providing a social security number; yet, the table design requires mandatory data entry in the Social Security data element. The existing design would be problematic, as it would prevent successful completion of patient registration. Again, the fit analysis between the task and technology enables identification of the issues and provides an opportunity for adjustment. In the case of purchasing a system, the opportunity for adjustments in design is limited, depending on the agreements already made and costs associated with changes. Ideally, the technology needs to respond to tasks and users' needs; however, in the real world, resources for customization are limited and we still need to figure out better ways to utilize the technology available at an affordable cost. The FITT analysis would still be helpful in this case. In order to increase fit, we can change one or both components. In the case when it is not possible to change the design, we can look into workflow or process adjustments, such that the task would still be accomplished successfully. If the fit is not taken into consideration, users will find the path of least resistance and create workarounds which may not be consistent and even create more issues in terms of effective, efficient, reliable and compliant task completion. Using the FITT analysis in such cases helps recognize the issue and provides an opportunity for effective and consistent interventions. Use of fit in this particular example would also be beneficial because it shows change in the task. Upon changes needed to address the task-technology fit, it may be necessary to revisit the individual-task fit and perhaps, provide additional task-related communication and training to users.

Some of the typical activities that occur during the implementation stage are development of the application/system, testing, system documentation, user training, and conversion from the old system to the new system. By using the FITT framework in this stage, there is still an opportunity to change the application while it is still being developed or tested, in order to improve the individual-technology and task-technology fits. For example, technology may impose a limit in the number of concurrent users, which could impact both individual-technology and task-technology fit. Again, FITT creates an opportunity to address the different components in relation to each-other. User trainings and system documentation can also contribute to improving the individual-technology fit. Results from the FITT analysis may impact the conversion alternative chosen. For example, if the analysis reveals a good fit between technology and users, as well as technology and tasks, there is a greater level of confidence in the system's performance, thus a direct cutover approach may be recommended. On the other side, if the fit between individuals and technology shows discrepancy when it comes to the level of preparation and readiness to use the system, a phased-in approach could be used; thus, allowing a little more time to adjust, learn, spread the knowledge, and gain more confidence in using the system.

Evaluation is the last stage of the systems development life cycle and includes activities such as backup, updates, upgrades, other maintenance aspects, and assistance for users. It is often at this stage that most organizations do an evaluation of the system's performance and may even survey users to find out their satisfaction levels with the new system. Retrospective analyses are good in terms of helping understand the process, what was done well, and what should be done differently next time; however, they are limited in terms of their potential for intervention and adjustment. The FITT framework can work in this stage, as well, especially when it comes to the individual-technology fit. Poor fits or gaps in user knowledge can still be addressed with further training and education. Updates and upgrades can also be based on the need for a better fit of technology with individuals or tasks. Overall, the nature of FITT evaluations and comparisons involved in the process makes FITT framework suitable to apply when the goal is to determe the difference between the aim and reality (Honekamp and Ostermann 2011).

## 16.6    Measurements Needed When Using FITT

In order to apply the FITT Model successfully, the three fits need to be measured. Ammenwerth et al. studied the case of a German university hospital through the use of FITT Model and used questionnaires to measure the three fits in the process of introducing a new nursing documentation system (Ammenwerth et al. 2006). Specifically, questionnaires and audit forms evaluated nurses' computer knowledge and attitude toward computers, nurses' acceptance of the nursing care process, nurses' satisfaction with the nursing documentation process, the quality of nursing documentation and overall effects of documentation systems on the workflow (Ammenwerth et al. 2006). Questionnaires that were used were standardized and validated. Documentation quality audits were also standard. Measurements were done before, during, and after the introduction of the new system, in three different units (wards) of the hospital. In addition, focused group interviews were conducted after the introduction of the system. Results of the questionnaires and audits determined whether individual-technology, individual-task, task-technology were a complicated fit or an uncomplicated fit. Based on the results, interventions were undertaken to address certain components, such as further computer training for users or clarification on nursing documentation expectations. Figure 16.2 is created based on some of the information from the German University Hospital case studied by Ammenwerth et al. (2006). Although, it does not represent the exact or complete research method and findings, it helps visualize the FITT evaluation process and serves as an example for applying the FITT framework in the active evaluation of a new information system.

While most healthcare organizations involve some of the users during the introduction and implementation of a new system, the input is not collected, analyzed, and considered in a structured manner. FITT framework provides somewhat of a holistic approach, as it prompts for a thorough review of the human resources (users), processes, and the technology in discussion. It also helps surface issues related to users and the technology. Application of FITT may be time-consuming, depending on whether the organization and the project team are well-versed on the new health information system, users involved (along with their knowledge and skill level), and the tasks affected. Knowing their employee skills and learning potential, having transparent, well-established processes and workflows, and having thorough knowledge of the information system they are about to adapt makes it easier to utilize the FITT model. When any of these components are not fully known, the FITT evaluation process may take longer; however, it is worth it as it makes the IT investment evaluation more rigorous and less likely to fail.

Another important note is that FITT is prone to constant change. External factors, such as updates in the technology, new hires, policy and procedure changes, process changes, differences among units, etc. create a dynamic situation when it comes to the three fit dimension, and thus, IT adoption process (Fig. 16.3). Also, certain changes are more difficult to address than others. For example, in the case of the German university hospital, the psychiatric unit was not satisfied with the task-technology fit. The system design made it complicated to document specific psychiatric information. On the other

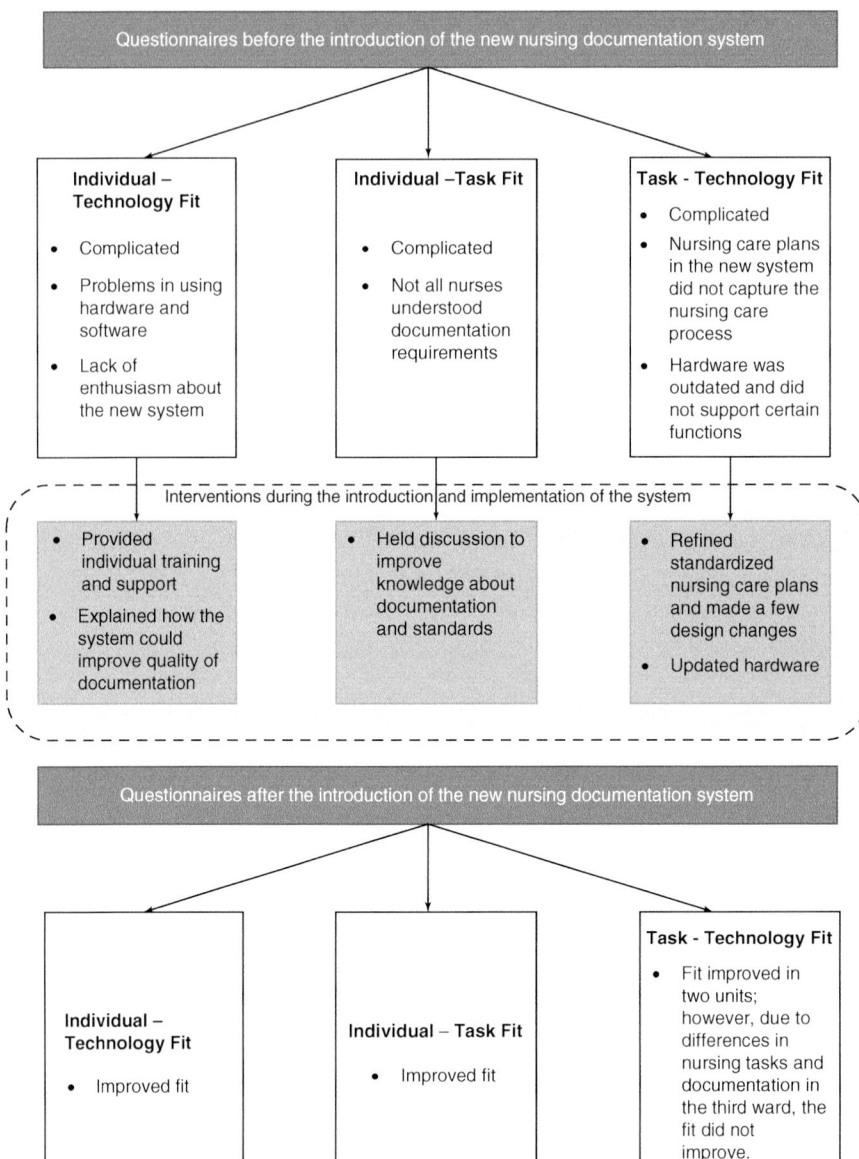

**Fig. 16.2** Evaluating a new nursing documentation system

side, changing the design to address the psychiatric documentation needs would mean distorting the design and the task-technology fit for the other two units. In cases like this, FITT analysis would be more beneficial if done during the planning and analysis phases (and not during the implementation phase). Nonetheless, FITT explains why some systems are successfully implemented in an organization but not in another.

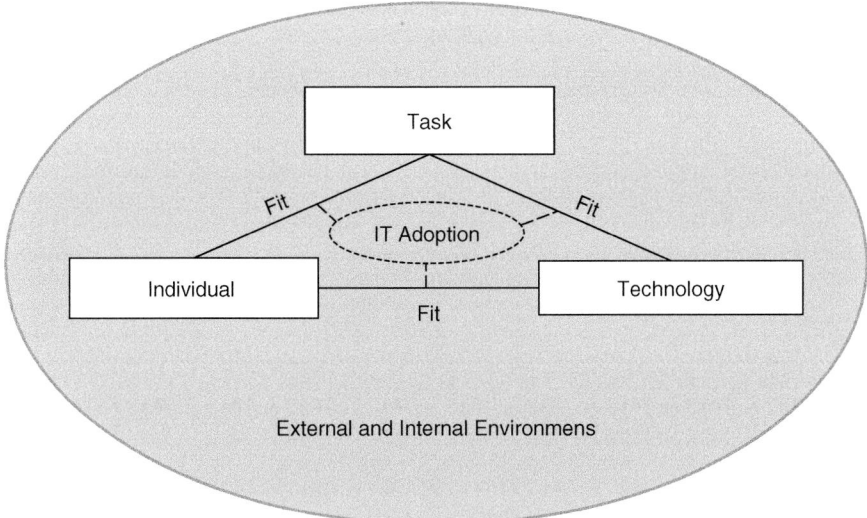

**Fig. 16.3** FITT Framework and the Environment (Based on Ammenwerth et al. 2006)

## 16.7 Closing/Final Thoughts

Health Information technologies, including mobile ones have expanded and continue to expand. As health care organizations and other parties continue to invest in them and count on them to support the efforts in improving quality, access, and cost of healthcare, it is very important to make sure that these new technologies are optimized. FITT model offers a tool that can help evaluate the process and outcomes of technology adaption, identify issues related to the fit between users, the tasks that need to be completed through technology and the technology itself.

## References

Ammenwerth E, Iller C, Mahler C. IT-adoption and the interaction of task, technology and individuals: a fit framework and a case study. BMC Med Inform Decis Mak. 2006;6(1):3–13. https://doi.org/10.1186/1472-6947-6-3.

Davis FD, Bagozzi RP, Warshaw PR. User acceptance of computer technology: a comparison of two. Manag Sci. 1989;35(8):982. http://search.proquest.com/docview/213229133?accountid=14872 Accessed 8 Dec 2017.

Goodhue DL, Thompson RL. Task-technology fit and individual performance. MIS Q. 1995;19(2):213–36.

Honekamp W, Ostermann H. Evaluation of a prototype health information system using the FITT framework. Inform Prim Care. 2011;19(1):47–9.

Krebs P, Duncan DT. Health app use among US mobile phone owners: a national survey. JMIR Mhealth Uhealth. 2015;3(4):e101. https://doi.org/10.2196/mhealth.4924.

Lesselroth BJ, Holahan PJ, Adams K, Sullivan ZZ, Church VL, Woods S, et al. Primary care provider perceptions and use of a novel medication reconciliation technology. Inform Prim Care. 2011;19(2):105–18.

Logan J. Electronic health information system implementation models – a review. In: Maeder AJ, Martin-Sanchez FJ, editors. Health informatics: building a healthcare future through trusted information; 2012. p. 117–23. https://books.google.com/books?hl=en&lr=&id=Px3vAgAAQ BAJ&oi=fnd&pg=PA117&dq=fit+between+individual,+technology,+and+task,+logan,+201 2&ots=GGS8lwuZDY&sig=s0RqH_7RYzmh3MA1IIAvtwHSFWM#v=onepage&q=fit%20 between%20individual%2C%20technology%2C%20and%20task%2C%20logan%2C%20 2012&f=false. Accessed 8 Dec 2017.

Markets and Markets. North American Healthcare IT Market by Product (EHR, RIS, PACS, VNA, CPOE, mHealth, Telehealth, Healthcare analytics, Supply Chain Management, Revenue Cycle, CRM, Claims Management) by end user (provider, payer) – Forecast to 2020. Oct 2015. http://www.marketsandmarkets.com/Market-Reports/north-america-healthcare-it-market-1190.html. Accessed 8 Dec 2017.

MedDataGroup. MedData Point on physicians adoption of mobile health. 2016. https://www.med-datagroup.com/infographic-mHeath-adoptions-and-challenge/. Accessed 8 Dec 2017.

Office of National Coordinator for Health Information Technology (ONC). Federal health information technology strategic plan 2011–2015. 2017. https://www.healthit.gov/sites/default/files/utility/final-federal-health-it-strategic-plan-0911.pdf. Accessed 8 Dec 2017.

Rogers EM. Diffusion of innovations. 5th ed. New York: Free Press; 2003.

Sayles NB. Health information management technology. An applied approach. 4th ed. Chicago: AHIMA Press; 2013.

# Chapter 17
# Portability

**Egondu R. Onyejekwe**

**Abstract**  The portability of a medical (or health) record assumes that it does exist and that it can be moved and accessed without distortions and/or damage. So, there must be laws that guide such activities. However, for today's portable health records, the laws fail to operate. Interoperability is not bypassed and electronic health records need to sustain privacy of the individual and the security of the content. For these among other reasons, they need to be commoditized to be portable. Two critical parts of the needed *revolution* process are the tools (the mobile gadgets and the human patient) that exist in today's portable markets. These should be made to off-set the hindrance of interoperability and content must remain fidelitous and or must be commoditized to be accessible through any mobile hand-held! Thus, the needed *revolution* for portable health records is partially here because the tools (portable gadgets and the patient—the walking server) are constants. All that is needed to complete the revolution is the commoditization of the "content" of the electronic health records. This will eventually bypass interoperability by applying these tools to create a truly secure and private portable health record!

**Keywords**  Portability · Electronic health record · Electronic medical record Portable health records · Portable medical records access · Interoperability HIPAA · HITECH · Disruptions

## 17.1  Introduction

The topic "Portability" which refers to "Portable Medical/Health Records" is imbued with many predetermined concepts or assumptions. The first is that health/medical records exist. The second is that not only can they be moved from one location to the other but, must do so with fidelity and to everyone who is qualified to receive them.

E. R. Onyejekwe (✉)
Public Health, Health Administration, College of Health Sciences, Walden University, Minneapolis, MN, USA
e-mail: Egondu.onyejekwe@mail.waldenu.edu

© Springer Nature Switzerland AG 2019                                                     209
E. R. Onyejekwe et al. (eds.), *Portable Health Records in a Mobile Society*,
Health Informatics, https://doi.org/10.1007/978-3-030-19937-1_17

Consequently, we will examine the forms and mechanisms that are currently used and that will be used to provide portability and universal access to persons' medical or health records. A critical discussion of a person's medical or health Record/s must therefore, in the United States' (U.S.'s) setting, be contextual and include: form; access; law and the limits of the law. So, the first question is, what is its form/nature and where will it reside? Secondly, who should have access and what mechanisms will be used for accessibility and portability? Thirdly, what is the role of the law and its limits? This third part, of course, relates to the US legal under-pinnings and the limitations of the free market, and so, begs the question of owner-ship. (But the law and its limits are discussed fully in a separate chapter.) Who owns a person's health/medical record/s, all be it a portable health record? Keep that thought! Below is a brief introduction of these three parts.

### 17.1.1   Form and Nature of Medical/Health Records

To be of critical value, a health record must be in an electronic form. It should be available and have value to both the provider and the patient, or their agent. If the health record is in electronic form it should, at a minimum, include various storage options such as cloud storage, mobile phone, tablet, flash drive, or any other storage medium in common use by today's society.

### 17.1.2   Access

For all practical purposes, access implies access to health records and by extension, access to informed healthcare. It is pertinent that the health/medical record is avail-able in a didactic exchange between the provider and the patient. Where then, would the health record reside to be so accessible? For example, could we securely store our medical records on Facebook and provide access to only authorized individu-als? How can we help alleviate real and perceived threats of public storage, while providing convenient access to both patient and provider? Could we create a time-based access key that would automatically expire in an established timeframe? Or could we possibly use advanced encryption techniques, such as a health record blockchain, to store our record securely and provide authorized access at the point of need? These all apply to the electronic form of the health or medical record.

Access can also be more broadly defined to include access to informed health care (regardless of the technology used to store the health record). The Penchansky and Thomas (1981) taxonomic definition of "Access" includes five dimensions. To date, these five dimensions of access (popularly known as the five As and discussed in a separate chapter of the book) are still relevant. The five dimensions are:

*Affordability*—addresses issues on insurance, including under-insurance and no insurance;

*Accessibility*—where care is delivered, including geography, transportation and mode;

*Accommodation*—How and when are patients seen, what sort of communications occur with the provider and what do the schedules look like;

*Availability*—basic supply and demand including mismatches as well as urban versus rural availability; and

*Acceptability*—mainly ethnic/cultural barriers and preferences.

These dimensions of access are not discrete but play crucial roles in concert with each other. In other words, they articulate the importance of each aspect and the interplay between the different aspects of access that reflect the fit between characteristics and expectations of the providers and the patients/clients. These must be understood in the context of the services provided and received. Consequently McLaughlin and Wyszewianski (2002) surmised that:

*Affordability*—becomes determined by how the provider's charges relate to the client's ability and willingness to pay for services;

*Availability*—measures the extent to which the provider has the requisite resources, such as personnel and technology, to meet the needs of the client;

*Accessibility*—refers to geographic accessibility, which is determined by how easily the client can physically reach the provider's location;

*Accommodation*—reflects the extent to which the provider's operation is organized in ways that meet the constraints and preferences of the client. Of great concern are hours of operation, handling of telephone communications, and the client's ability to receive care without prior appointments.

And finally, *acceptability* captures the extent to which the client is comfortable with the more immutable characteristics of the provider and vice versa. These characteristics include the age, sex, social class, and ethnicity of the provider (and of the client), as well as the diagnosis and type of coverage of the client.

McLaughlin and Wyszewianski (2002) argue that the "five As of access form a chain that is no stronger than its weakest link." For example, if affordability is improved by providing health insurance but the other four dimensions have not also been improved, this does not imply any significant improvement of access and utilization. Often neglected are the basic characteristics—including socioeconomic status, age, gender and other differential statistics of the provider and the client that influence acceptability. Likewise, equating access with the availability of resources misses other provider/patient characteristics that may be barriers to access. Variations in the use of healthcare services may emanate from regional differences, including the heterogeneity amongst patients in different locations that induce or debar access. So, the mere presence of facilities is not an adequate measure of availability, because it misses a more important issue of goodness of fit—the interaction between the characteristics of the providers and the expectations of the patients that invariably determine how acceptable the resources are to the patients. Maybe, a more reliable measure of the goodness of fit between provider and patient is for a patient to have a regular physician and a regular site of care, where all the patient's electronic health records can be accessed at the point of care. This, by itself, then minimizes the need for portability since it reflects access in almost all its forms, from availability, accessibility, accommodation, to acceptability. All it would need then would be affordability. Given this affordability gap, the full picture on access still does not emerge because affordability plays a role in influencing utilization. Instead, disparities occur. There is a growing body of research that points to racial and ethnic differences in the utilization of different medical, dental and other services. These other nonfinancial predictors, particularly, point to the critical role played by all of the dimensions of access, but they especially point to availability, accessibility, and acceptability.

In the context of the present discourse, how does the electronic health record meet the requirements for affordability, accessibility, accommodation, availability and acceptability?

### 17.1.3   The U.S. Health and Healthcare Laws

In the United States, there are federal, and state administered, laws that govern health and health care. They deal with issues related to the United States health care system, covering a vast array of legal subjects ranging from affordable health care, emergency health care, health insurance, health reform, mental health to patient protection. This chapter deals with only three of those laws at a Federal level.

First, is the Health Insurance Portability and Accountability (HIPAA) Act of 1996. This is a United States legislation that provides data privacy and security provisions for safeguarding medical information. The Office for Civil Rights is charged with enforcing the HIPAA Privacy Rule and the Security Rule. HIPAA Privacy Rule protects the privacy of individually identifiable health information. The HIPAA Security Rule sets national standards for the security of electronic protected health information.

What if and when the ownership of electronic health records is distributed? Can HIPAA regulations be supported in electronic health records that are no longer owned by a specific entity? Or will there be new rules for distributed liability when the health records are collectively owned? Would the Federal government, the state government, independent organizations, the patient or even the patient's agent be responsible for insuring the privacy and security of the patient's electronic record? Or would an electronic record broker be mandated to take care of the issues of privacy and security?

Next is the *Health Information Technology* for Economic and Clinical Health (HITECH) *Act* of 2009. The HITECH Act provides Health and Human Services (HHS) the authority to establish programs to improve health care quality, safety, and efficiency through the promotion of health IT, including electronic health records among others.

Again, how will this HITECH Act be adjusted to accommodate the lack of a single defined owner of electronic health records? This must be addressed when the electronic health record is distributed and is collectively owned.

The third is the Affordable Care Act, (ACA) passed by Congress and signed into law by the President in March 2010. ACA gives individuals better health security by establishing comprehensive health insurance reforms. These insurance reforms hold insurance companies accountable, lower health care costs, guarantee more choices, and enhance the quality of care for all Americans. Even, the ACA would need to be modified by a portable and collectively owned electronic health record.

Summarily, portability would over-ride the single and dominant ownership of electronic health records. Yet, it would encompass a form of the health record that can be accessed and that operates within the confines of the stipulates of some form of US health and healthcare related laws that address distributed/collective ownership. Presently, the form and nature of the electronic health record is overwhelming for many people. It requires us to take a close look at how electronic health records are currently used and managed.

The rest of this chapter discusses access in its broader terms, and the laws along with their implications. The next section of the chapter will focus on assessments of electronic health records; their forms and how access is affected by the laws. The final section will be devoted to Disruptions, particularly those attributes such as interoperability, portability, and invariably. These would call for a need for change even in the current form of electronic health records (EHRs). Furthermore, current laws would need to be adjusted to accommodate (both allow and support) legal records in the portable electronic format/s. So, here is a look at electronic health records.

## 17.2 Electronic Health Records

"An *electronic health record* (*EHR*), or *electronic medical record* (*EMR*), is the systematized collection of patient and population electronically-stored health information in a digital format (Gunter and Terry 2005)." The contents and or composition of EHRs vary. They range from simple demographics to vital signs, medical history, medications, or laboratory test results. They may include allergies, immunization status, radiology images, personal statistics like age and weight, as well as billing information. Besides the contents, EHRS can be shared in and among a variety of healthcare settings, through network-connected, enterprise-wide information systems or external and other information networks and exchanges.

Often, the terms EHR, electronic patient record (EPR) and electronic medical record (EMR) are used interchangeably, although differences between models are now being defined by the industry. For all practical purposes, EHR, EPR and (EMR) are used interchangeably. However, EHR relates to a more longitudinal and inclusive collection of electronic health information of individual patients. Contrast this with the electronic medical record (EMR), which is the patient record created by providers for specific encounters in hospitals and ambulatory care environments. EMR can thus serve as a data source for an EHR (Habib 2010; Kierkegaard 2011).

At the other end of the spectrum is the personal health record (PHR). The federal government's healthcare IT web site, HealthIT.gov, summarizes the PHR as an electronic application for recording personal medical data that the individual patient controls and may make available to health providers.

EHR is the only definition adopted in this book as it is inclusive of EMR. EHR is also the term adopted by the industry and practically, all medical practices in the United States have adopted electronic health record systems (either as EHRs or EMRs). There is a far-ranging array of EHR vendors that provide systems that are certified to have the features and functions specified by the Office of the National Coordinator for Health Information Technology (ONC). Given this wide array of systems, it is very difficult for practitioners (especially physicians) to choose a specific EHR vendor.

A major deterrent and an attribute of current EHRs is the hard link processing—where the programs are designed specifically to work with a given proprietary record format. That means that entities not associated with the proprietor would be unable to access the EHRs. There has to be a way to commoditize the health record systems. Invariably, this implies the need for a means/mechanism to break the data away from the programs that are used to feed a proprietary model. Commoditizing the health record format, makes it widely available and interchangeable with another such record format. This provides standardization of the health records that will enable systems to collaborate, share, and exchange information.

An essential attribute of an EHR system is to accurately store the data and capture the state of a patient across time. Consequently, it should eliminate the need to track down a patient's previous paper medical records and assists in ensuring data is accurate and legible. It should therefore, reduce risk of multiplicity and or the replication of patient data since there should be only one modifiable version. Given the health record format is commoditized, the concept of a file for records would be eliminated and (if the record is stored in a distributed ledger with millions of other such records) access would be seamless. This also, implies that the record is more likely to be up to date and decreases risk of lost data, as well as reduces the risks associated with a patient who moves from location to location and from clinician to clinician.

Undeniably, the influence of the government (the laws plus incentives) has rapidly moved the market for electronic health records to a more mature stance for a section of the market. A major driving force is heralded by the influence of government incentives through the American Recovery and Reinvestment Act of 2009 that offered incentives for purchasing and using EMR systems in a meaningful way. They offered extra payment to physicians and hospitals who were early adopters and who, therefore, prioritized purchasing of EHR faster than they would have otherwise. The act also ascribed penalties for physicians who submitted paper claims to Medicaid or Medicare.

None of these still addresses portability of the electronic health record. Invariable, a portable EHR should be independent of time and place! A portable EHR, must be available to the patient whose data it records. It should be capable of holding the aggregate or all of the patient's health record and must have the ability to be searched and arranged based on the needs of the patient at the time of an encounter with a healthcare provider (physician or other), independent of temporal and spatial constraints. This is the fundamental theme of this book, that EHRs should be available, accessible, indisputable, usable, impregnable, and PORTABLE. In other words, the patient should be able to access their EHR from wherever they are and whenever they need it, regardless of where it is stored and or the device used to access it.

The notion of portability of the EHRs potentiates a digital format that is aggregated, searchable, interchangeable, and so forth for the patient. It also potentiates the ability to extract medical data within and across patients that relate to possible disease/condition trends as well as long term changes in the patient/s. Ultimately, commoditized EHR formats will further and accelerate population-based studies in healthcare.

While mergers and acquisitions are well known to the health IT market, a seemingly emerging strategy is attempting to reach a greater number of healthcare organizations, providers and patients through synergistic pairing. The Trump administration foreshadows a heavy regulatory burden facing healthcare professionals with more regulatory changes ahead.

Joseph G. Cramer, M.D., a Utah paediatrician was sure his practice did due diligence in choosing an EMR. However, at the end, he found out that it was all wrong! This excerpt from his perspective is very revealing.

> Please, please, please do not do what we did. Do not purchase the wrong electronic medical record (EMR) system. This advice comes from someone who has championed the virtues of EMRs for decades. The difference is that I went from theory to reality. And our clinic's reality stinks (Cramer 2010).

The problem wasn't that they didn't try. On the contrary they tried extremely hard. They were convinced of completing their due diligence of evaluating all things technical. They constituted an EMR committee from the single specialty pediatric clinics in the Salt Lake City and Park City, Utah, metropolitan area, that met for months in order to sort through 'the morass of vendors.' The vendor list included more than 400 variations that even confounded things much more for them. Neither did the attendance to many and a variety of tech conferences offered by their specialty society and national trade shows help. From the investigation of the companies, testing the software, and hearing the sales pitches to having many more committee meetings, it all came to naught. He asserted:

> And we all lost. The hurt continues to this day, more than 2 years after the contract was signed. I share our blunders so that you don't repeat them. Consider it a gift from a friend who still believes in the power of data and the desperate need for doctors to have the right tools to manage information and practice quality medicine. I am so frustrated, however, that my anger gives me the strength to write (Cramer 2010).

## 17.3   Disruptions: A Failure to Focus on Design

There are so many ways to assess disruptions with regards to EHR. The failure to focus on design was one aspect of their failure according to Joseph G. Cramer. However, how Joseph G. Cramer interpreted their failure is a necessary but an insufficient part of their failure. To him, the failure to focus *on the most important part of the decision—the human/computer interface* was instrumental to their failure. The host of subjects addressed included: practice management system handed claims; the viability of the vendor; and reliance on national standards and certification. Even their IT representative approved of their computer language. But, having failed to listen to their *guts* regarding the design of the computer screen, which invariably they would be looking at for several hours a day, he opined, was their demise. He further articulated that "it is about functionality and workflow. It is all about design, which we see every day, but mostly ignore…. Design of the computer screen and the underlying program is how our brains see the whole picture of the patient."

He notes that "Human/computer interface is the key to success in adopting an EHR (Cramer 2010)." Proper design leads to doing the right thing. Poor design makes it easy to do the wrong thing—the door doesn't open, or it slams shut on your fingers. The mistake is attempting to recreate a paper document on the computer instead of making a usable display. It is as though the paper chart still reigns supreme.

So, it is not enough to know that there are vendors who provide EHR software services, it is equally crucial that physicians (clinicians or other) who make these choices are able to add some valuable context within the decision-making process. For example, having a comprehensive view of the EHR vendor offerings and the ability to discriminate amongst them may result in a compatible choice for the practice or it may not result in a compatible choice for the practice! Both options must be available to them. A rather profound point he articulated is regarding the conceptual view of the EHR, purely as another view of the record that is documented on paper.

Specialized Knowledge and Applications (SK&A) is a Healthcare marketing company that provides healthcare reference information. SKA's reports (from the calls they made to healthcare practitioners between September 2016 and February 2017) are packed with lots of data that provide granular level insights on EHR data. Physician EHR usage rates are provided, but are further broken down by practice specialty, region, states, practice size and patient volume. While all of these may be beneficial to Healthcare Information Technology (HIT) enthusiasts, groups and companies, as well as stakeholders, a basic question remains: can physicians (practitioners) capture this granularity for an efficient practice?

Furthermore, the Utah pediatrician, Joseph G. Cramer, MD opined that better presentations of EHRs by vendors or others do not automatically imply better EHR products. Rather, practitioners should choose EHRs that adapt to their processes and not fix their processes towards the dictates of the EHR (Cramer 2010).

## 17.4    Disruptions: Interoperability

Even if the first entry for EHR is into the free market place, there are perturbations that will likely disrupt this marketplace. At the core of this disruptive potential is health record portability or movement. In today's systems, electronic health records are not portable because they fail to talk (communicate with each other). They cannot communicate with each other because the electronic health record format is not commoditized. It still operates more or less as a static file system that is distributed in private servers and that are hampered by *interoperability*. What is interoperability?

HealthIT.gov defines interoperability of health information thus:

According to section 4003 of the 21st Century Cures Act, the term 'interoperability,' with respect to health information technology, means such health information technology that—" "(A) enables the secure exchange of electronic health information with, and use of electronic health information from, other health information technology without special effort on the part of the user; " (B) allows for complete access, exchange, and use of all electronically accessible health information for authorized use under applicable State or Federal law; and " "(C) does not constitute information blocking as defined in section 3022(a)(Office of the National Coordinator for Health Information Technology 2018)".

Interoperability, therefore, describes the extent to which systems and devices can exchange data, as well as interpret the shared data. That means that if two systems are interoperable, they are not only able to exchange data, but also capable of presenting that data so a user can understand it. This implies that a health record that is ported from one system to another must remain Fidel or remain as the same health record at the other end. But this is not the case, because, the current state of the disparate healthcare systems does not meet this interoperability requirement.

There are several barriers that mar the benefits that could be derived from health information exchange, whether it is an electronic health record or other. Among them are those that satisfy data protection as they move (or are ported) from one system to the other. Protection includes both the privacy of the individual and the security of the data contained therein. Additionally, are those that would not change the contents of the data within the record. Even as new tools emerge, there is an insufficiency of interoperability tools to handle the porting of data efficiently and effectively. At HIMSS17, one additional issue that needed to be resolved was that all stakeholders shared the importance that when data is exchanged, it is done in a way that protects the privacy of individuals as well as meets compliance requirements.

So, by 2017, interoperability or this critical ability to share patient data electronically had not been achieved. A section of HIMSS17 focused on "The Interoperability Imperative: Value-based Care Depends on Health Information Exchange" (Kalorama Information 2018) where new approaches were being applied towards solving this long-standing interoperability challenge. The approaches included both emerging standards and protocols, as well as the innovative use of existing technologies. Multiple presentations showcased—emanating from pre-conference symposia to the Interoperability Showcase at HIMSS17—were charged with moving the interoperability bar forward and towards a more connected future state. Multiple presentations and demonstrations focused on emerging standards, protocols, APIs and apps, as well as the innovative use of existing technologies. The amalgamation of new models for information exchange, which included the use of data analytics, availed HIMSS17 attendees with creative new solutions worth considering. Addressed also, were current obstacles, challenges and evolving models, with a goal of fully realizing solutions in such areas as patient engagement, trusted exchange, and image sharing.

As the U.S. moves to the "value-based" model of care, achieving interoperability becomes imperative. Attempts are underway to at least, leverage the existing regional and statewide Health Information Exchange (HIE) to address the opioid crisis which has diminished the returns on U.S investments in the healthcare industry. HIE is also being leveraged for streamlining regional access to a single patient portal. Additionally, are other innovations that potentiate a more connected future state for the U.S. Health care. Amongst them are mining HIE data to realize predictive analytics or incorporating state immunization registry reporting to automate statewide exchange.

Furthermore, there are other solutions that are addressing the need for patient engagement, trusted exchange, and image sharing. Among such future-focused ideas is the potential use of *blockchain* to address secure health record sharing. Future healthcare interoperability efforts may rely increasingly on blockchain. (*Blockchain is a technology developed to support virtual currency bitcoin*. It is discussed at length elsewhere in the book). HIMSS17 featured an all-day event that explored the use of blockchain. One of the sessions "Blockchain in Healthcare: *A Rock Stars of Technology Event,*" featured Tamara StClaire, a health IT consultant, who urged HIMSS17 attendees to look past the hype claims surrounding blockchain and to grasp its real potential.

"A lot of people believe there's going to be momentum and acceleration," said StClaire, adding that blockchain could "impact almost every healthcare transaction" and "has a global ability to change the way we think (Milliard 2017)."

There were also discussions "of a developing collaboration that will leverage standards-based exchange for the 2020 Tokyo Olympics." Michael Nusbaum Board Member HIMSS, elaborated from the show floor on "how the Olympic Healthcare Interoperability Initiative (OHI) will demonstrate the benefits of interoperability at the 2020 Olympic games in Tokyo."

But numerous barriers still prevent full health information exchange (HIE) among providers. The Executive Director of the California Association of Health Information Exchanges, in "Breaking HIE Barriers," opined that HIE is evolving beyond the basic data exchange. This requires agreement on transport standards and authorization standards. Interoperability that adds content standards and vocabulary standards toward "access" where the detailed specification of and access is fashioned only to what you want to know. However, he cited these three barriers: providers expect HIE to be easy and ubiquitous; providers and patients aren't interested in just data movement; and that HIE hasn't demonstrated its value!

For all the focus on interoperability technology and standards, HIMSS17 attendees were reminded that interoperability" or enabling easier sharing of electronic health records is a means to an end, which remains—improving patient-centered care. Interoperability is all about *portability*—enabling the exchange of electronic health record (information) with the goal of improving clinical collaboration and patient care. John Gresham, Vice President, Cerner, wraps it all up by stating: "Interoperability leads to a more complete clinical record, which can be used to create accurate, actionable insights for decisions regarding patients' care." EHRs cannot be ported without interoperability!

Interoperability therefore, allows for portability of EHRs. Interoperability subsumes the commoditizing of the EHR begins by creating a seamless access process that eliminates the concept of a file for records and one that has the record stored in some form of a distributed ledger with millions of other such records. This also, implies that the record is more likely to be up to date, and decreases risk of lost data, as well as reduces the risks associated with a mobile patient whose geographical location changes frequently and who is in continuous motion from provider to provider.

A patient in continuous motion thus loses the utility value of their EHR! Unlike everything else a person owns (which gets classified as their property), the electronic health records are not so categorized or defined. So far, healthcare costs continue to rocket, and even where practitioners such as physicians and hospitals object, their threats are geared towards switching vendors and not towards redirecting the ownership of EHRs or towards making them accessible.

Patients (clients) are beginning to clamor more because there is still tremendous value in EHR usability, vendor-switching, and lack of mindshare in the market. Healthcare still involves many local markets and then with this explosion of handheld tools, social media and the like, EHR is a type of software with tremendous potential to get directed and entrusted in the hands of the rightful owner, the patient/client. Portable devices and the web provide great opportunities to sell/locate patient EHR information directly where it is used—where both the provider and the patient (client) can directly access it as needed.

Yet, there is a lot of competition, with hundreds of scattered companies serving local or web markets, and a few large entities that are capturing large institutions. There have been government involvements, from stipulating privacy and security laws to funded physician and hospital incentives in the largest market of the United States. These transformed to potential penalties, for noncompliance. Even with the increased depth of physician and hospital adoptions, a variety of dissatisfied clientele emerged. There is user dissatisfaction with their purchases because of information blocking, regulations, consumer technology growth, lack of interoperability, implementation difficulties, staffing shortages in IT, lack of a clear EHR market leader, increasing international markets, and cyber security. All these mar the progress towards portable health records.

Providers also wish to have unhindered patient EHR access during a didactic engagement with the patient (client). Additionally, the web and the social media provide newer and more opportunities on how differential access mechanisms could be provided to both the patient and the provider. These are trends that are combining to disrupt the EHR marketplace, even while revenues for companies grow. So, stay tuned as the healthcare we know may or may not transition to a "value-based system with more patient involvement."

## 17.5 Positive Signs and the Way Forward

Today's EMR/EHR is no closer to *portability,* than yesterday's was. Neither is focused on supporting a portable health record. Today's EHR is predicated on vendors who wish to corner the market on EHR. These vendors goals are not really geared towards the creation of any portable EHR ecosystem. As a consequence, the current US EHR ecosystem operates in a health care environment that is riddled with inefficiencies and high costs of delivery. It has no defined route/s for Access (in all its dimensions). There are limited ways to attest for the accuracy of the information in the record. The EHR itself is also, hindered by various laws and legal underpinnings!

So, the way forward must be *revolutionary*! Towards this end, I have borrowed from different sources that include blogs at Zobreus.com (Zobreus 2015).

My first source opined that a person's medical record is his or her health "story." Therefore, the person should be involved in it, and that it should not be left entirely in the hands of providers or vendors. The person involved should be able to tell their health story (Zobreus 2015)! Studies of patient-oriented care including one cited by Health Affairs (James 2013) show that a patient who is more engaged (active) in shaping their story, is a better care recipient. Patients engaged in "shared decision making" with their providers, yield two positives, lower costs and improve patient outcomes. This is a process that allows the patient to help in shaping their medical/health record, and consequently getting the best possible care. Providers, including doctors, should encourage more patients to read their medical records. A byproduct of reading their records, is that doctors may be more inspired to write more thoughtful and accurate notes, given that their patients would likely read them. After all, the Health Insurance Portability and Accountability Act of 1996 accorded patients the rights to access their medical records, and the right to electronic copies since 2009. Yet, most patients never see their health records or charts (Khullar 2016)!

No patient should be passive about their medical/health records! They should let their doctors know of their interests in both accessing their records, as well as in helping in narrating their story. After all is said and done: "Your medical record is ∗*your*∗ story, you should tell it (Zobreus 2015)."

The Universal Vaccination Card (Zobreus 2015) bypasses the use of loose leaves of paper for the documentation of immunization. The health records here are not commoditized. Rather, the company has built a technology, including their award-winning interface that allows any clinic to easily upload a patient's shot data to a centralized portal. With that, those vaccination records are instantly available to the patient on their mobile device. The concept is rather clear because any authorized entity such as camp, school, government body, etc. can access the person's complete, digital shot card instantly from the company's portal. Alternatively, the patient can provide it from their device. Those shot records are "official" because they are entered by person's providers (and verified by the company). Because the person's shot card is universal and comprehensive, *it moves with the person* regardless of where they go including when they switch providers. No currently existing practice—specific shot record can travel with the patient or person. This provides a true sense of portability (at least for this part—the immunization—of the electronic health record).

Here is a more generic question "why does having your medical record matter?" One especially poignant answer came from Dr. Dhruv Khullar in his insightful *New York Times* blog titled, "Let Patients Read Their Medical Records (Khullar 2016)." In actuality, there are several compelling reasons—enough to warrant a separate post. Suffice to say that medical records have gone past the paper folders that were static in book shelves. Health records in those paper formats hardly exist in the US anymore for several reasons. Among them are the terrible waste including taking up space and not being eco-friendly—resulting in high healthcare costs and the danger they pose for violating HIPAA. Among the several dangers they are vulnerable to are: they could be destroyed by natural or man-made disasters; they can be stolen; or they can be misplaced or lost. These are just a few reasons why health records, have gone digital and as of this writing, the electronic health or medical record is the backbone of the entire U.S. health system. But, there are issues and concerns regarding these electronic health types of records. One is that they are merely conversions from paper copies to electronic copies. (This issue will be revisited shortly.) Another problem (not necessarily encountered with paper versions), is the ease with which doctors can apply the copy and paste function for the information in a medical record. This, for sure has notable advantages. For example, forwarding text for stable patients can be safe and efficient—and many doctors do not think it hurts patients. However, if not carefully executed, it can result in at least two major problems.

The first is error and its propagations in the patient's record. The other is that it is a can be a potent source of "note bloat" i.e. the notes so filled with extraneous information that the provider has have to scroll through "pages and pages of nonsense to find anything useful." A third problem can be associated with cognition. First a doctor's experiences for rummaging and reviewing the huge amount of auto-populated information is different from where the doctor simply searches for, confirms and records aspects of a patient's health history. So, when a doctor reads a patient's electronic health record, s/he will likely assume that "what's written there is as likely to be wrong or outdated as it is to be accurate." While such discrepancies can sometimes be minor and inconsequential; they can sometimes be devastating and are relevant to life or death underpinnings.

What can get lost in all this is the patient's story. Doctors should be skilled enough to elicit, distill and communicate an account of what's happened in their patient's life. Furthermore, "gathering and sharing a patient's story offers the fullest sense of who a patient is as a human being, why he might have received this treatment, for example, and not that one, and what the best course of action might be going forward. We now spend 2 h a day reporting quality measures, but what needs to be mandatory in the age of digitalization is the art of story gathering and storytelling."

## 17.6  Conclusion

There are no quick answers to portability. So, to answer the question: "why does having your medical record matter?" Or even more specific "why mobile?" The "default" answer which is easy to overlook in modern times is that a person's "record already *is*, and always has been, "mobile (Zobreus 2015)." This is evidenced in many developing parts of the world as well as in the early days of medicine in the advanced world. That means that your medical record already resides with you, be that in your head, via your vital signs, on some scratched notes in your pocket, etc. However, and as medical information became more voluminous, it became more "efficient" for providers to maintain your written records instead. But then, disruptions ensued as health care further specialized, and people started to move from city-to-city, and practice-to-practice. The inefficiencies of paper records became obvious. This heralded the electronic medical record era! Unbeknown to us is the fact that "what we now have is simply a migration of your scattered record from practice-specific paper charts, to practice-specific electronic ones." Consequently, there has not been any meaningful *interoperability* gain!

In the interim, nearly 80 percent of today's world population carries a mobile "server" either as hand-held or in their pockets. This is a technology that will allow us to securely have our records at all times. That means that our health records must bypass the interoperability hurdle or be commoditized to be accessible through the mobile hand-held.

Next is the patient, the single most important *constant* at all health encounters, and thus "a walking server." The patient, the "walking server" already has the "mobile server" and both mitigate against medical record file fragmentation, be these in paper or electronic formats! This union—of distributed health record ledger—is inarguably the most efficient solution to medical record fragmentation! There's no longer any efficiency from having your records scattered in different silos—one major advantage of a commoditized record. Neither the providers nor the patient gets benefit/s from silo-based health records, which also comprises patients as well (Zobreus 2015).

Getting back to the "default," that people *are* walking "records" we content that people are imbued with their own histories. They know their bodies and were the ones that go/went to the appointments as well as, the ones that have/had the ailments. But, while the current U.S health system is already inefficient at accurately getting this information from us, those little servers in our hands or in our pockets are not. The mobile devices have a lot of potential to deliver precious information from us back into the health system, and not simply the other way around. They can start anywhere, from fit tracking data, to spotting changes in behavior, to taking photos, or to even tracking adherence. This comprehensive electronic medical record tool is an all-around winner and becomes priceless because we can control it, and it could/can capture and process all of this information to, and from, the distributed healthcare system ledger as well. Perhaps even an imperative… is this: So, the question really should be "why *not* mobile (Zobreus 2015)?"

This chapter therefore, concludes that the *revolution* we need is partially here because the tools we need are here. Those tools include the mobile gadgets and human patient who act as the "mobile-servers." All we need is the second part of the *revolution*—commoditization of the contents of the tools—the EHR, to eventually bypass interoperability and create a truly portable health record!

# References

Cramer JG. How to choose the 'right' electronic medical record system. 2010. http://medicaleconomics.modernmedicine.com/medical-economics/news/modernmedicine/modern-medicine-feature-articles/how-choose-right-electronic-m?page=full. Accessed 27 Jan 2018.

Gunter TD, Terry NP. The emergence of National Electronic Health Record Architectures in the United States and Australia: models, costs, and questions. J Med Internet Res. 2005;7(1):e3. https://doi.org/10.2196/jmir.7.1.e3.

Habib JL. EHRs, meaningful use, and a model EMR. Drug Benefit Trends. 2010;22(4):99–101.

James J. Health policy brief: patient engagement. Health Aff. 2013. https://doi.org/10.1377/hpb20130214.898775/full/

Kalorama Information. EMR market 2017: round-up of HIMSS 2017. 2018. https://www.kaloramainformation.com/Content/Blog/2017/02/27/Round-Up-of-HIMSS-2017. Accessed 27 Jan 2018.

Khullar D. Let patients read their medical records. 2016. https://well.blogs.nytimes.com/2016/03/31/let-patients-read-their-medical-records/. Accessed 27 Jan 2018.

Kierkegaard P. Electronic health record: wiring Europe's healthcare. Comput Law Secur Rev. 2011;27(5):503–15. https://doi.org/10.1016/j.clsr.2011.07.013.

McLaughlin CG, Wyszewianski L. Access to care: remembering old lessons. Health Serv Res. 2002;37(6):1441–3. https://doi.org/10.1111/1475-6773.12171.

Milliard M. Blockchain's potential use cases for healthcare: hype or reality? 2017. http://www.healthcareitnews.com/news/blockchains-potential-use-cases-healthcare-hype-or-reality. Accessed 27 Jan 2018.

Office of the National Coordinator for Health Information Technology. Interoperability. 2018. https://www.healthit.gov/topic/interoperability. Accessed 27 Jan 2018.

Penchansky R, Thomas JW. The concept of access: definition and relationship to consumer satisfaction. Med Care. 1981;19(2):127–40.

Zobreus. Your medical record is your story, tell it. 2015. https://www.zobreus.com/20150321yourmedical-record-is-your-story-tell-it/. Accessed 27 Jan 2018.

# Part VI
# Impact

# Chapter 18
# Impact

**Dasantila Sherifi**

**Abstract** Despite the emergence of electronic health records and patient portals, given the multiple points of data collection and storage, accessing complete health information on a timely manner can be a daunting task. Incomplete and delayed access to health information leads to delayed care and may impact health outcomes of individuals and populations. The expansion of health information exchanges (HIEs) across the US helps with sharing of information; however, there are challenges related to patient identification, adherence to health information privacy, and other technical aspects. Technologies such as SaaS-based Master Patient Index solutions, crowdsourcing, cloud computing and other web-based technologies can help address some of these challenges.

## 18.1 Introduction

Consider this scenario. Jane, a 51 year old woman has her annual checkup with her family doctor. Among others, the doctor follows up with Jane on her cholesterol medication, the mammography and colonoscopy she was supposed to have had during the past year. Jane reports compliance with medications, thanks to pharmacy reminder calls for refills. She reports that she just had the mammography done a month ago because she had forgotten until then. The doctor is able to access the radiology report and sees a note that the right breast images were not clear, and therefore, Jane needs to do another mammography of the right breast. There is also a note that a call was placed to notify Jane about this but Jane did not receive the call. She only answers calls on her cell phone, not home phone. Last, Jane did not do the colonoscopy because she lost the doctor's "paper." The doctor tells Jane that she could have called or e-mailed her through the portal to receive the colonoscopy order or even ask for the order to be sent directly to the GI doctor. Jane responds that

D. Sherifi (✉)
College of Health Sciences, DeVry University, Naperville, IL, USA
e-mail: dsherifi@devry.edu

© Springer Nature Switzerland AG 2019
E. R. Onyejekwe et al. (eds.), *Portable Health Records in a Mobile Society*,
Health Informatics, https://doi.org/10.1007/978-3-030-19937-1_18

bled and accessed as needed from a distributed network of storage locations. This sounds like a great vision for creating an integrated complete and up-to-date medical record that would include every record that is generated for the patient from any healthcare providers, as well as from patients themselves. From a technical perspective, this model requires that all providers have 100% electronic health records and fully interoperable systems. In addition, it is important that multiple users are only able to edit or delete the records they have generated themselves. Each entity contributing to the distributed network environment should be able to still maintain some type of ownership over the information they have created and contributed to this shared environment. From a legal perspective, it requires for all input and output to be authenticated, date and time stamped, and categorized into different levels. For example, should a hospital be asked to produce the legal health record for Jane Doe, the hospital should be able to retrieve and produce all of the records and only the records that have been generated by that hospital. Should a physician office be asked to release certain documents from a patient record, they should be able to only retrieve those specific documents; otherwise, there is a risk for invasion of privacy. From a user perspective, there may be challenges when it comes to understanding and interpreting the information correctly.

Imagine a well-designed distributed network environment where information comes from physician offices, hospitals, lab services, pharmacies, patients, etc. In an effort to increase patient engagement, the system allows the patient to sign up for instant updates. Every time new information is added, patient is notified. Depending on the education level, health record design and format, as well as ability to analyze the new information in the context of other information, patients may not understand or misunderstand. There is also potential for loading incorrect information. It is not uncommon to create duplicates, overlaps or overlays when entering new patient information. Correct patient identification is critical when information is shared instantly. Misinformation may create potential for additional unnecessary health services (in that case, resources are spent on clarifying the situation rather than providing care for the patient) or present a health information breach risk for the healthcare organization. SaaS-based MPI solutions can help address these patient identification issues and quality control processes can help address the accuracy of information but they would require additional work, collaboration, and agreements among the various healthcare providers.

Crowdsourcing could be another model that supports access to integrated health records. Crowdsourcing is used successfully to support freelancers and other professional communities. One of the challenges with crowdsourcing may be the varied level of understanding between providers and patients, as well as reliance on the open platform. This could be a risky platform considering sensitivity of health information and various privacy and security regulations.

While web-based and mobile technologies show much promise for the future of healthcare information and patient engagement, some efforts need to remain focused on improving online access to health record. A survey of 502 consumers conducted by HealthMine (2016) in 2016 found that 53% of the respondents do not have online access to their healthcare information. In addition, 32% had difficulty accessing their health information, 29% had difficulty accessing lab results, 29% had difficulty accessing their health insurance information, and 25% had difficulties accessing their prescription history. Access issues need to be addressed at all levels in order for the above discussed vision to work.

## 18.3  Cloud

Cloud computing is already used by healthcare organizations when it comes to archiving data or providing application software on multiple servers. Can we continue to leverage cloud based resources in an effort to store and manage integrated health records? With the expected increase in the amount of data collected by healthcare organizations and individuals, as well as increase in the number of web-based applications, the need for cloud computing will increase. Dedicated, reliable hosts are highly available and they offer many advantages in comparison to a local physical computer or server. Andrzejak, Kondo, and Anderson point out that cloud-based services provide a greater level of security and more efficient disaster recovery than local servers would provide (Andrzejak et al. 2018). It provides a better platform to facilitate increased collaboration that would result from increased communication between patients and providers. The cloud provides automatic software updates and document updates. There is also potential for increased reliability in documentation because existing documents can be updated without a need to replicate them; although this last characteristic should be used carefully, given certain rules and regulations that apply to healthcare documentation. For example, should an incorrect entry make it into a health record, the provider is not allowed to destroy the original submission. Instead, the provider would be asked to create an amendment and clarify the correct entry in the health record. Such concerns have already been addressed and resolved, and healthcare organizations are using cloud-based storage effectively. Cloud computing coupled with better integration with relevant databases can be leveraged to positively identify patients, avoid redundancy and increase the accuracy of the patient record. Non-dedicated resources may be explored in the future but at this time, they are much more volatile, less reliable, and more prone to security breaches, which makes them undesirable for use in storing and sharing healthcare information.

Mobile technologies are also supported by cloud-based services. Cloud servers may be accessed over long-range communication and mobile users are able to input data or retrieve data as output. Mobile hardware is typically at a disadvantage when compared to static hardware. The use of mobile applications increases, more work is being done to explore the use of cloud services as sources of computational power and energy (Li et al. 2016). As we leverage these types of resources, the capability of mobile devices will increase, and thus, enable consumers to increase their use for healthcare purposes.

# References

Ammenwerth E, Lannig S, Horbst A, Muller G, Schnell-Inderst P. Adult patient access to electronic health records (Protocol). 2017. https://www.cochranelibrary.com/cdsr/doi/10.1002/14651858. CD012707/epdf/full. Accessed 19 Jan 2018.

Andrzejak A, Kondo D, Anderson D. Exploiting non-dedicated resources for cloud computing. 2018. http://mescal.imag.fr/membres/derrick.kondo/pubs/andrzejak_noms10.pdf. Accessed 19 Jan 2018.

Health Information Management Systems Society (HIMSS). Mobile technology upgrades continue to impact healthcare engagement. 2015. http://www.himss.org/news/mobile-technology-upgrades-continue-impact-healthcare-engagement-0. Accessed 19 Jan 2018.

HealthMine. Survey: patients lack online access to health records. Inf Manag J. 2016;50(3):17.

Li C, Yanpei L, Youlong L. Efficient service selection approach for mobile devices in mobile cloud. J Supercomput. 2016;72(6):2197–220.

Schwartz PH, Caine K, Alpert SA, Meslin EM, Carroll AE, Tierney WM. Patient preferences in controlling access to their electronic health records: a prospective cohort study in primary care. J Gen Intern Med. 2015;1(30):25–30.

# Chapter 19
# Communication

**Allison Chinyere Nnaka**

**Abstract** The portable health record (PHR) is recommended as modern resolution to the problems of fragmented communication and absence of inter-operability amongst diverse electronic medical record (EMR) systems. Portable health record allows the primary care provider to share health information about a patient with other health care professionals and institutions, including specialists, laboratories, and nursing homes to improve the safety and quality of health care, especially during emergency care. Portable health record support patient centered healthcare by making medical records and other relevant information accessible to patients, therefore assisting patients in health self-management. Patients can use PHRs in one of three formats: a provider-maintained digital summary of clinical information accessible to patients; a patient-owned software program that allows users to view and update their own health information; or portable, interoperable digital files with which patients can manage and transfer information. PHR also enable patients to refill prescriptions, access lab results, track immunizations, and schedule appointments. Electronic exchange of health information raises questions about policies and procedures concerning privacy, security, and identity management. Many health providers are unenthusiastic to give up discretion of their records, and many electronic medical record vendors have found the method of generating multifaceted processes to change one record to another to be expensive and time consuming. It is recommended that the application of existing legal and privacy provisions should be continuously addressed as PHRs develop. Because PHRs are developing health information technologies, the legal and privacy concerns concerning their use may change as technologies and their roles in health information technology more largely evolve.

**Keywords** Benefit of PHR · Providers and patient communication · Provider to provider communication · Barriers and recommendations

A. C. Nnaka (✉)
Public Health, Health Administration, College of Health Sciences, Walden University, Minneapolis, MN, USA
e-mail: Allison.nnaka@dc.gov

© Springer Nature Switzerland AG 2019
E. R. Onyejekwe et al. (eds.), *Portable Health Records in a Mobile Society*, Health Informatics, https://doi.org/10.1007/978-3-030-19937-1_19

## 19.1   Introduction

The use of electronic records has been widely known as an efficient way to improve the provision of health care and enable health care providers to access and share patient information. Health care providers may document patient's medical history in number of ways (Clarke et al. 2006). Electronic medical records (EMRs), for illustration, are digital copies of patient charts commonly used in physicians' offices to record patient data. Electronic health records (EHRs) are more comprehensive in scope, including information from all the clinicians involved in a patient's treatment, such as immunizations, family medical histories, and previous providers (Clarke et al. 2006). Primary care physicians can share EHRs with other health care professionals and institutions, including specialists, laboratories, and nursing homes. Personal health records are documented in personal health records (PHRs). However, unlike electronic health records, which are only accessibly to clinicians, patients can use personal health records to manage and update their own medical information (Kaelber et al. 2008). PHRs empower patients to take control of their health record, improve their health status and improve clinical outcomes, because they help patients monitor health conditions and effectively communicate with health care providers. Patients use personal health records in one of three formats: a provider-maintained digital summary of clinical information accessible to patients; a patient-owned software program that allows users to view and update their own health information; or portable, interoperable digital files with which patients can manage and transfer information (Bickford 2015). PHRs in mobile format (mPHRs) fall into the third category and allow patients to access health information through the Internet or telecommunication devices, such as cellular phones (specifically, smartphones, or cellular phones that includes an operating system capable of running general-purpose applications and performing many of the computer functions), personal digital assistants, and tablet computers.

Personal health records and the electronic health records have served patients well in their interactions with health care professionals. Consumers and health care practitioners thus far have used personal health records largely in nonemergency settings (Chen and Zhong 2012). In times of emergency, sick or injured individuals may be displaced or not in a condition to provide their personal health records. This creates serious challenges in post-disaster care (Chen and Zhong 2012). In such situations, reliable sources of clinical information are invaluable to patients who cannot communicate or receive treatment from caregivers who are unfamiliar with their medical histories (Chen and Zhong 2012). Given the challenges associated with communicating during disasters, integrating personal health records and mobile personal health records into emergency response plans could help ensure quality health care delivery if or when existing methods of information sharing (e.g., paper or computer-based records) fail. Moreover, the growth of self-management tools for remote monitoring, particularly those available in Web and mobile application formats, contributes to the increased use of PHRs and mPHRs among consumers.

## 19.2    Benefit of Personal Health Record

The increasing use of mPHRs among patients reflects a broader trend in health care digitization; the growing popularity and utility of mobile medical applications. Such applications function on the above-mentioned mobile devices, which have rapidly evolved into ubiquitous tools for sharing information and communicating with others (American Psychological Association 2012). Mobile devices also have the ability to withstand certain types of infrastructural failures during disasters. As such, they may be uniquely qualified to play important roles in responding to public health emergencies (PHEs). Personal health record tools can improve clinical outcomes. For example the use of personal health records has been associated with improved self-monitoring and positive clinical outcomes for hypertension, adherence to immunizations and other practices supporting child wellness, and management of medications (Bickford 2015). However, the benefits of personal health records in improving clinical outcomes may be correlated with age, because younger patients are more likely to use personal health records frequently.

## 19.3    Use of Portable Health Record in Providers and Patient Communication

Just as the shift from paper-based records to EHRs provides numerous benefits to providers and patients, mobile personal records certainly provide more benefits to patients than other types of personal health records. Increasing and widespread Internet access and mobile device use allow patients to access their records from anywhere with an Internet connection (Ball et al. 2007). Mobile personal health records offer providers a method to share information with patients, including clinical summaries, diagnoses, educational resources, and appointment reminders (Ball et al. 2007). They also enable patients to refill prescriptions, access lab results, track immunizations, and schedule appointments. Some of these features (e.g., prescription refills) exist in applications developed by major retail outlets such as Walgreens and CVS, but do not comprise a holistic health record. However, other mobile services (e.g., Group Health, Castlight Mobile, MyChart, myCigna, Coventry Mobile, MHBPSM Mobile, Evita Personal Health Record, and Capzule PHR) do serve as holistic records of health information. Other applications such as Health4Me enable patients to view their insurance claims, track health spending, and search for local health care professionals.

The functions of PHRs have the potential to create a more broad and well-adjusted understanding of the patient, because patients can control and manage the information in the record. Personal health record therefore permit patients to note relevant medical information anytime and share information, such as emergency department visits and other unprepared visits with providers (Garfunkel Wild 2011). The PHRs also enable continuity of care if a patient obtains treatment from a different provider. The clinician in this type of situation would have access to a detailed record of the patient's existing medical conditions, including previous medical tests, procedures, prescriptions, and conditions (Bickford 2015). Such information, in turn, would prevent duplication of tests and treatments, as well as minimize the risk of administering medications that can complicate conditions or allergies.

There are several mPHRs that serve the abovementioned functions, some are particularly custom-built for emergencies. An example of such custom built personal health records is the Microsoft HealthVault, which allows users to generate medical records for unanticipated emergency or hospital visits or to inform first responders (Ball et al. 2007). The Gazelle application, allows smartphone users to receive and share their lab results with providers. The Web-based PHR service known as the Synchart stores patients' health information and can grant clinicians emergency access to information during emergency. Mobile personal health records provides health care providers immediate access to patient's medical history and new medical events that can be helpful in both emergency and nonemergency situations (Garfunkel Wild 2011). During public health emergencies, when many patients are displaced or health care facilities lose abilities to access EHRs, mobile health records is one of the ways of providing accurate, current medical information.

When patients are unable to seek care from their primary care physician or a facility with access to their medical history, mPHRs can help providers obtain essential information, such as medical conditions and drug allergies, needed to determine treatment options and better coordinate and direct care (Conn 2018). Such information could be useful when health systems are overwrought and facilities may lack satisfactory numbers of staff. Mobile personal record can notify providers of important health information that can eventually reduce medical error and improve triage.

Mobile health records also help certain populations during public health emergency, such as nonresponsive patients who cannot communicate with providers, those who seek care at another care facility, susceptible and distinct populations, and children and young adults (Clarke et al. 2006). For vulnerable and special populations, such as non-English-speaking persons, those belonging to ethnic minority groups, mentally unstable patients, and those who are deaf or blind, mobile personal record may be the only communication method between patient and provider (Bickford 2015). Additionally, mPHRs afford a method for young population to become answerable for decisions concerning their care and develop self-sufficiency, especially for those who do not have access to routine medical care (Cox 2013). mPHRs also inspire parents to be more involved in preventive medical care for their children; for example, parental PHR use is linked to enhanced immunization adherence.

## 19.4   Use of Portable Health Record in Providers Communication

The delivery of health care through telecommunication technology relies on methods such as real-time videoconferencing to facilitate medical care in emergency scenarios. Telehealth strategies, videoconferencing, the use of smartphones and wireless networks, and email have proven to be effective at sharing information between clinicians and medical facilities. PHRs also enable information exchange between qualified clinicians, and can assimilate communication and virtual imaging capabilities (Garfunkel Wild 2011). Therefore, they offer a platform for virtual visits, permitting office-based health care providers and home-based health care workers or patients to coordinate patient care management. Similar basic patient information, such as blood pressure interpretations, temperature, glucose levels results, and other medical notes, can be conveyed from home providers to physicians via a personal health record (Garfunkel Wild 2011). These functionalities empower physicians to remotely obtain patient histories and educate care providers at home the essential changes to the patient's plan of care, which can be a significant step toward structuring home telehealth capabilities (Clarke et al. 2006). PHRs have the potential to enable better communication between providers, families, caseworkers, and others making care decisions on behalf of the sick person.

## 19.5   Challenges and Barriers to Use of Portable Health Records

In spite of the likely advantages of integrating mPHRs into typical medical practice and disaster response efforts, some significant challenges linger. Portable health record are becoming increasingly complicated. From a technical perspective, PHRs may allow interactive communication between patients and providers, and the integration of PHRs and EHRs would permit exporting data among information systems (Clarke et al. 2006). An underlying challenge is that PHRs alone do not have universal standards and there is no one standardized way to design and maintain PHRs. Even if such challenges were resolved, and despite the potential benefits of integrating PHRs and EHRs, several factors would still inhibit their integration. First, it remains unclear how health system roles and responsibilities will change if systems are integrated. For example, concerns about liability risk and adverse effects for providers, such as increased workload and inadequate reimbursement, remain unresolved (Teodecki 2010). Second, there is an absence of standards to inform the process by which systems should be integrated. Furthermore, it is not clear whether there are limitations to the current health information technology (HIT) infrastructure that could present technical challenges to integration. Third, related to limitations of current infrastructure, concerns about privacy, security, the use of information by third parties. On a different note, ethnic and racial minorities have been reported to adopt personal health records less frequently than ethnic and racial majorities do. Also, patients from lower income groups are less likely to use personal health records as compared to those with higher incomes.

Because PHRs and mPHRs are evolving health information technologies, the legal and privacy issues concerning their use may change as health information technologies generally evolve (Pew Internet Research Project 2012). Select PHRs, offered by health care providers and health plans, are covered by the Health Insurance Portability and Accountability Act of 1996 (HIPAA) Privacy Rule. In the event that HIPAA applies to a PHR or mPHR, the information within these records is protected by the law. However, those systems that are not offered by HIPAA covered entities must adhere to the privacy policies and their respective vendors. A system that is not covered by HIPAA may be covered by other applicable laws; however, all PHR and mPHR system providers should recognize how health information is protected and convey such policies to patients.

## 19.6 Recommendations

Electronic and mobile personal health records have the potential to empower patients through greater access to personal data, health information, and communications tools, which may aid self-care, shared decision making, and clinical outcomes. They may increase patient safety through exposing diagnostic or drug errors, recording non-prescribed medicines or treatments, or increasing the accessibility of test results or drug alerts (Adams and Corrigan 2003). They may also reduce geographical barriers to patient care and act as a point of record integration, particularly in fragmented health systems, thus improving continuity of care and efficiency.

Although the majority of PHR development in the United States takes place in the private sector, the federal government is best suited to design, implement, and regulate PHR use, given its involvement in emergency preparedness and response efforts at the national, state, regional, and local levels. The federal government also oversees health care delivery and innovation in HIT and is therefore well positioned to implement the appropriate standards for PHR standardization and interoperability. Congress has a duty to identify criteria to regulate the use of mPHRs and allow qualified providers to receive inducement payments, established on the standards set for EHRs (U.S. Congress HealthIT 2014). The American Recovery and Reinvestment Act of 2009 has authorized incentivized payments for providers to encourage adoption and use of EHRs, such legislation may perhaps authorize incentive payments to providers that meet criteria for the use of mPHRs (U.S. Congress

HealthIT 2014). Legislation needs to emphasize the possible use of mPHRs during emergencies. Given that many private corporations develop and monitor mPHRs, establishing criteria for how providers and health care systems use and access these records can maximize the benefits of mPHRs to clinical outcome.

## 19.7   Conclusion

The changing landscape of the health care industry, from the development of new forms of health coverage, to the adoption of new legislation, create a system that is much different from health care two decades ago. The industry has adopted technology, as a means of making health care more affordable and efficient (Deloitte 2018). New forms of reimbursement, such as those through managed care systems, require oversight on all sides, including insurers, physicians, and patients. Patients are typically the ones who make decision about which types of coverage and physician offices they prefer. At the same time, the development of electronic methods for communication of health information allow physicians and practices to communicate more easily with insurance companies and third-party payers. Technology also allows patients and practitioners to communicate at a moment's notice. Whether the increase in the number of insured will make electronic communication more difficult or more valuable to a practice and patients, will depend on how technology savvy the patients and practitioners are and whether practitioners will use the technology to remain organized in tracking health communications with other providers and patients.

## References

Adams K, Corrigan JM, editors. Priority areas for national action: transforming health care quality. Washington, DC: National Academies Press; 2003. p. 1–13.

American Psychological Association. The advantages of electronic health records. 2012. http://www.apa.org/monitor/2012/05/electronic-records.aspx. Accessed 14 May 2018.

Ball MJ, Smith C, Bakalar RS. Personal health records: empowering consumers. J Healthc Inf Manag. 2007;21(1):76–86.

Bickford CJ. The specialty of nursing informatics: new scope and standards guide practice. Comput Inform Nurs. 2015;33(4):129–31.

Chen T, Zhong S. Emergency access authorization for personally controlled online health care data. J Med Syst. 2012;36(1):291–300.

Clarke JL, Meiris DC, Nash DB. Electronic personal health records come of age. Am J Med Qual. 2006;21(suppl 3):5–15S.

Conn J. Modern healthcare: no longer a novelty, medical apps are increasingly valuable to clinicians and patients. 2018. http://www.modernhealthcare.com/article. Accessed 14 May 2018.

Cox P. World faces mounting damage from disasters. 2013. http://www.aljazeera.com/indepth/features/2013/05/20135278951818557.html. Accessed 14 May 2018.

Deloitte. The mobile personal health record: technology-enabled self-care. 2018. http://www.deloitte.com. Accessed 14 May 2018.

## 20.2 The Need for Coding

Codes existed even before computers were invented, and all computer systems must utilize codes for their operation (Browne 2008). Most of us will encounter medical coding at some point in our lives. Whether as a professional practitioner, insurance biller, patient, family member of a patient, or insurance customer service representative, all of these stakeholders see these codes or use these codes whenever a procedure is performed or treatment is conducted as part of ongoing services provided in the healthcare industry. Unfortunately, many different coding systems have developed throughout the years. It is very important that the meaning of these codes be accurate and transparent within each system vertically and across the board horizontally among all of the different coding systems. This concept constitutes the fundamentals of semantic interoperability.

Technological advancement in health informatics and biomedical engineering has paved way to various Health Information Systems (HIS) in the healthcare industry. Even though these information systems along with the associated medical equipment have contributed to positive health outcomes, the most important issue still remains concerning how patient data is used. The information created may originate from various sources, so it may not even be saved in a unified database within the same hospital. In many instances, a single hospital may have a Radiological Information System (RIS), a Laboratory Information Management System (LIMS), and other HISs which are not even connected to one another (Kolias et al. 2014).

So that these problems can be addressed, standards such as Electronic Health Record (EHR) and Health Level 7 (HL7) have been created (Kolias et al. 2014). However, these two standards still have a considerable limitation in that they both do not possess the semantic information of medical information in a configuration that can be easily identifiable and processed by computers. Because of this, medical data is unfortunately concealed in various data pools.

In order to solve this problem, semantic web technologies can be used to furnish the tools that enable healthcare workers to process information much more effectively and accurately. Semantic web technologies can also construct the framework for interoperability among all of the different HIS. They can also help integrate information from different sources by defining the semantic meaning.

In the past few years, many coding systems were developed to introduce terminologies in healthcare in order to integrate medical information. Some examples of these systems are International Classification of Disease (ICD), Current Procedural Terminology (CPT), Logical Observation Identifiers Names and Codes (LOINC), Systematized Nomenclature of Medicine Clinical Terms (SNOMED CT), Healthcare Common Procedure Coding System (HCPCS), RxNorm, RadLex, Unified Code for Units of Measure (UCUM), Foundational Model of Anatomy (FMA), Medical Dictionary for Regulatory Activities (MedDRA), and Unified Medical Language System (UMLS).

## 20.3 Coding Standards

There are many different coding standards presently being used in the healthcare industry. It is very beneficial for all stakeholders to know the purpose and usefulness of each of these different coding standards. As mentioned in the previous section, they are as follows: ICD-10, CPT, HCPCS, SNOMED CT, LOINC, RxNorm, RadLex, UCUM, FMA, MedDRA, and UMLS. Every one of these coding systems will now be described in detail.

## 20.4 ICD-10 Coding Standard

The International Classification of Disease Version 10 (ICD-10) is an internationally recognized coding standard developed by the World Health Organization (WHO). ICD-10 utilizes alphanumeric codes representing a diagnosis, symptom, and even cause of death (HealthFusion 2014). ICD-10 is widely used throughout the world and utilizes an convention that is universally agreed upon. This coding system ensures that a health professional in one country will be able to interpret a diagnosis in the same exact way as health professional in another country. ICD-9 was used as one of coding standards in the United States prior to October 1, 2015, but ICD-10 has since taken over ICD-9 since that date. The latest 2018 ICD-10 codes became effective on October 1, 2017, and all claims made in 2018 must use these updated 2018 codes (ICD10Data 2018). The main difference in these ICD versions is that ICD-10 has longer and more detailed codes than ICD-9. The longer and detailed formats solved some of the issues that had been encountered in the previous ICD version.

## 20.5 CPT Coding Standard

Current Procedural Terminology (CPT) is a coding system used in the United States developed by the American Medical Association (AMA). CPT focuses on the services provided by the medical office or hospital, and it is used widely by insurance firms to reimburse physicians for diagnostic and therapeutic services that had been provided.

## 20.6 LOINC Coding Standard

Logical Observation Identifiers Names and Codes (LOINC) is a coding system that had been devised by the Regenstrief Institute in 1994. LOINC is a universal standard for clinical and lab test observations, and it can facilitate the exchange of information among different coding systems. In contrast to LOINC which codes for testing and observations, ICD-10 records diagnosis, and CPT records clinical service (HealthFusion 2014).

## 20.7 SNOMED CT Coding Standard

The Systematized Nomenclature of Medicine Clinical Terms (SNOMED CT) is a coding system that provides a common footing for different medical offices and hospitals. It is regarded as the most comprehensive, multilingual coding system. The codes in SNOMED CT can be easily mapped to other coding systems such as ICD-10, and SNOMED CT is an excellent example of semantic interoperability at work. SNOMED CT is used to record patient data such as patient medical, family, and social histories (HealthFusion 2014).

## 20.8 HCPCS Coding Standard

The Healthcare Common Procedure Coding System (HCPCS) is the coding system utilized by the Centers for Medicare and Medicaid Services (CMS). This coding system ensures that Medicare and Medicaid among other medical insurance claims are handled in a systematic and uniform way. Originally, HCPCS was used voluntarily, however, with the advent of the Health Insurance Portability and Accountability Act (HIPAA), the use of HCPCS became compulsory (Centers for Medicare and Medicaid Services 2017).

## 20.9 RxNorm Coding Standard

RxNorm is the entire electronic catalog of standard nomenclature for pharmaceuticals and also drug delivery equipment used in the United States. RxNorm enables semantic interoperability for electronic systems that deal with drugs and drug delivery devices. This coding system facilitates lucid communication between electronic systems no matter the software or hardware used (TechTarget 2017).

## 20.10 RadLex Coding Standard

RadLex is a complete radiology lexicon used in standardizing all radiology terminology for practice, education, and research (National Institutes of Health 2016). RadLex consolidates and supplements other electronic standards such as SNOMED CT and LOINC. Although RadLex is comprehensive, this radiology coding system is constantly adding new terminology to be inclusive of technological advances, which fills critical gaps.

## 20.11   UCUM Coding Standard

Just like LOINC, the Unified Code for Units of Measure (UCUM) was also devised by the Regenstrief Institute. UCUM is a unified coding system for units of measure to be communicated between humans and electronic systems and also from electronic systems to electronic systems. This coding system is used to describe laboratory tests, clinical examinations, pharmacetical data (German Institute of Medical Documentation and Information 2016). UCUM can derive all units of measure from seven basic building blocks of fundamental units. These fundamental units of measure are meter, second, gram, radian, Kelvin, Coulomb, and candela (German Institute of Medical Documentation and Information 2016).

## 20.12   FMA Coding Standard

The Foundational Model of Anatomy (FMA) was designed by the Structural Informatics Group from the University of Washington. It was created to describe human anatomical structures. The FMA coding system consists of more than 75,000 different anatomical types (Kolias et al. 2014). These types range from parts that are sub-cellular to parts that are primary to the human anatomy. The 75,000 anatomical types are integrally tied in with 130,000 other anatomical terms that symbolize commonly used terms, synonyms, and even non-English translations (Kolias et al. 2014). FMA delineates and illustrates 2.1 million anatomical relationships as well.

## 20.13   MedDRA Coding Standard

The Medical Dictionary for Regulatory Activities (MedDRA) system is an immensely specific and structured collection of terminologies describing regulatory information for medical products utilized by humans. MedDRA covers products such as vaccines, biologics, pharmaceuticals, and drug-device combination systems (Zenuni et al. 2015). It is very useful in the monitoring the safety before and after the sale of a medical product.

## 20.14   UMLS Coding Standard

The Unified Medical Language System (UMLS) was designed by the U.S. National Library of Medicine (NLM) to help in the development of electronic systems to think in the language of health and medicine. There are three UMLS Knowledge Sources: the Metathesaurus, the Semantic Network and the SPECIALIST Lexicon. The Metathesaurus is a database constructed by meanings and concepts. It contains source vocabularies from coding standards such as ICD-10, CPT, SNOMED CT,

LOINC, and RxNorm. The Semantic Network is comprised of broad subject categories, which are known as semantic types. Built into the Semantic Network are also very practical relationships between semantic types, which are known as semantic relations. The SPECIALIST Lexicon is a database consisting of a general English dictionary with embedded health and medical terminology (U.S. National Library of Medicine: Fact sheet 2013).

## 20.15    Semantic Interoperability Enabling Clinical Data Discovery

There are various promising new projects underway that is looking to merge the rich clinical information from electronic health records with the corresponding genetic sequencing data. Patient data from across the United States would be grouped into large patient cohorts, and each patient within each cohort would provide samples documenting his or her genetic sequencing (Wells 2016).

The stakeholders associated with these promising projects come from all different backgrounds. They are industry leaders, private foundation entrepreneurs, and officials from the United States government. Their ambitious intentions are quite exemplary, and the huge potential for finding something innovative deserves the time and effort expended on these initiatives. However, there are semantic interoperability conditions and concerns that have to be investigated and ascertained before work in done on the combination of these targeted clinical data.

If these standards are not set, then the merging and alignment of these rich clinical information will only produce a limited amount of useful data that is of value. For example, being able to identify genetic determinants of a malady or disorder from this extensive population, heavily depends on clinical data established on standards in a way that sizable subpopulations can be inter-compared utilizing agreed upon clinical terminologies and codes.

The standards to be discussed can be grouped into one of two classifications. The first category deals with the currently used coding systems. The second category deals with the critical value lists that have to be delineated.

The first category incorporates well-defined coding systems that have already been discussed in the previous sections of this chapter. Examples of this first category are RxNorm codes for medication, LOINC codes for lab orders and results, ICD-10 diagnostic and procedure codes, HCPCS codes for CMS, and CPT procedure codes. The CPT, ICD-9/ICD-10 coding systems have been utilized in electronic health records and billing systems for quite some time now. LOINC and RxNorm are relatively new when compared to CPT and ICD-9/ICD-10.

The second category deals with data pertaining to very important patient information that is utilized to subdivide each patient cohort. For example, they are patient data such as height, weight, and blood pressure which are used to determine patient vital signs. Patient allergies, history of tobacco use, age, gender, race, and socioeconomic status, are some of the other examples of Category 2.

Category 2 is the most difficult to delineate due to the shortage of prevailing standards within the United States. This implies that the private companies and the federal government initiating the conception and formulation of these patient cohorts have to delineate standards beforehand and arrange for funding to the providers who will allow their clinical data to be mapped to these standards.

In order to prevent an overlap of data or duplicate counting, some kind of patient identification and matching algorithm must be used in order to make sure that clinical data from the same patient going to different providers are associated with the same patient in the cohort. In addition, it is imperative all examination reports, pathology reports, and diagnostic reports can be read and analyzed by agreed upon software to extract clinical data from the compilation and assortment of unstructured text in these reports.

There are many challenges ahead for these patient data initiatives. Discoveries from such an enormous patient cohort's genetic sequencing data can only happen if stakeholders ensure that semantic interoperability is in place to make sure that invariable genetic profiling of millions of people in the patient cohort occurs (Wells 2016).

## 20.16  Standards for Apps

The extent of the mobile health's (mHealth's) grasp in the healthcare industry is projected to grow to 16.4 billion US dollars by the end of 2018, and it is forecasted to hit $49 billion US dollars by 2020 (Peterson et al. 2015). mHealth is indeed as real and American as apple pie. Increasing numbers of people are using smart devices like smart phones and smart watches because of popular and practical mHealth apps. Apps are even being developed by patients and their families to meet increasing demands and needs in the mHealth market. With their increasing popularity and use, mHealth apps must conform to and comply with the standards and laws that govern them. All apps must comply with the Health Information Portability and Accountability Act (HIPAA). It is imperative that all privacy issues be clarified and addressed before any app is allowed to be used. Just as important, there is an absolute need to identify implementation standards (Peterson et al. 2015) for these apps and for any future inventions in the ever-changing field of mHealth.

## 20.17  Conclusion

In summary, semantics is study of what things mean in communication. In the context of electronic health records, semantics refers to the meaning of codes utilized by people within the healthcare community. These codes are a necessary to represent medical procedures and treatment and to facilitate billing. Codes existed even before computers were invented, and all computer systems must utilize codes for their operation. Whether as a professional practitioner, insurance biller, patient, family member of a patient, or insurance customer service representative, all of these stakeholders see these codes or use these codes whenever a procedure is performed or treatment is conducted as part of ongoing services provided in the healthcare industry. Unfortunately, many different coding systems have developed throughout the years. It is very important that the meaning of these codes be accurate and transparent within each system vertically and across the board horizontally among all of the different coding systems. This concept constitutes the fundamentals of semantic interoperability. In the medical community, semantic interoperability is the ability of different coding systems to recognize each other in order to provide a seamless transition within the healthcare industry. It is very important that coding used in electronic health records do its job of communicating comprehensive and accurate meaning of the health information that it represents. There are many coding standards currently in the healthcare industry. This chapter covered useful coding systems such as ICD-10, CPT, HCPCS, SNOMED CT, LOINC, RxNorm, RadLex, UCUM, and UMLS. Semantics and semantic interoperability are crucial in portable healthcare as they facilitate advances in the use of electronic health records.

# References

Browne E. Codes, EHRs and semantic interoperability. 2008. https://openehr.atlassian.net/wiki/spaces/healthmod/pages/2949134/Codes+EHRs+and+Semantic+Interoperability. Accessed 4 Nov 2017.

Centers for Medicare & Medicaid Services. HCPCS-general information. 2017. https://www.cms.gov/Medicare/Coding/MedHCPCSGenInfo/index.html. Accessed 4 Nov 2017.

German Institute of Medical Documentation and Information. UCUM. 2016. https://www.dimdi.de/static/en/klassi/ucum/index.htm. Accessed 4 Nov 2017.

HealthFusion. Coding standards: what's the difference between ICD, CPT, LOINC and SNOMED CT? 2014. https://www.healthfusion.com/blog/2014/health-topics/medical-coding/whats-difference-icd-cpt-loinc-snomed-ct/. Accessed 4 Nov 2017.

ICD10Data. The web's free 2018 ICD-10-CM/PCS medical coding reference. 2018. https://www.icd10data.com/. Accessed 28 May 2018.

Kolias VD, Stoitsis J, Golemati S, Nikita KS. Utilizing sematic web technologies in healthcare. In: Koutsouris DD, Lazakidou A, editors. Concepts and trends in healthcare information systems. Annals of information systems, vol. 16. Cham: Springer; 2014.

National Institutes of Health. RadLex. 2016. https://www.healthdata.gov/dataset/radlex. Accessed 4 Nov 2017.

Peterson C, Adams SA, DeMuro PR. mHealth: don't forget all the stakeholders in the business case. Medicine 2.0. 2015;4(2):e4. https://doi.org/10.2196/med20.4349.

TechTarget. RxNorm. 2017. http://searchhealthit.techtarget.com/definition/RxNorm. Accessed 4 Nov 2017.

U.S. National Library of Medicine: Fact sheet. Unified medical language system. 2013. https://www.nlm.nih.gov/pubs/factsheets/umls.html. Accessed 4 Nov 2017.

Wells B. Calling for semantic interoperability standards that enable clinical data discovery. 2016. http://www.healthcareitnews.com/blog/calling-semantic-interoperability-standards-enable-clinical-data-discovery. Accessed 14 May 2018.

Zenuni X, Raufi B, Ismaili F, Ajdari J. State of the art of semantic web for healthcare. Procedia Soc Behav Sci. 2015;195:1990–8. https://doi.org/10.1016/j.sbspro.2015.06.213.

# Chapter 21
# Analytics

**Dasantila Sherifi**

**Abstract** Big Data is the term for a collection of data sets so large and complex that it becomes difficult to process using on-hand database management tools or traditional data processing applications. With thousands of petabytes of data created every day, vast opportunities are created for data analysis and creation of new knowledge. The industry has responded to this opportunity by developing various data analytics tools, such as Hadoop. Big Data and Big Data Analytics tools have already been proven invaluable for many industries in the process of sales, decision-making, market segmentation, and customization of products and services. Healthcare has also started to enjoy the benefits of Big Data and analytics in areas such as personalized care, improved efficiency and effectiveness, and research. The expansion of mobile technologies in healthcare promises to make data collection and data usage more convenient. The expansion of consumer healthcare applications has created new opportunities to add even more to the existing pool of Big Data; thus, new opportunities emerge for data analytics. This chapter will provide an overview of Big Data and Big Data Analytics tools and some ideas for further development and enhancement of such tools.

**Keywords** Big data · Big data analytics · Healthcare analytics · Healthcare data · Data analytics · Portable health records · Mobile health analytics

## 21.1 Introduction

A few decades ago, statistician and quality guru, W. Edwards Deming, and management guru, Peter Drucker, both highlighted the importance of data in the management process with their saying, "You can't manage what you don't measure". The thousands of petabytes of data collected by various industries (including healthcare) that are carried over the internet each day are an indication that we have become well advanced in our ability to collect data. In fact, these ginormous amounts of

D. Sherifi (✉)
College of Health Sciences, DeVry University, Naperville, IL, USA
e-mail: dsherifi@devry.edu

© Springer Nature Switzerland AG 2019                                            249
E. R. Onyejekwe et al. (eds.), *Portable Health Records in a Mobile Society*,
Health Informatics, https://doi.org/10.1007/978-3-030-19937-1_21

data, called "Big Data" have become very complex in healthcare. They present opportunities and challenges in terms of measuring and using them in a meaningful and beneficial way.

## 21.2 Importance of Data Analytics

Big Data Analytics provide excellent opportunities for improvements in healthcare in terms of disease treatment and management, as well as disease prevention for individuals and population health (White House Big Data Report 2014). Individual-specific data combined with genomics and medication data can drive personalized medicine to a higher level which can improve medication safety and efficacy. The data maintained by electronic health records for each individual patient creates opportunities for better understanding of the individual health over a larger period of time, including physiological and pathophysiological changes and interelations among different body systems (Belle et al. 2015). The combination of such detailed clinical data with other nonclinical data can be very powerful in assessing, treating, and preventing disease. From a population health perspective, data analytics enables important indicators such as timely monitoring of disease outbreaks, incidence of certain diseases, prevalence of diseases within communities, or rates of vaccination. Such indicators are used to address issues, reallocate healthcare resources and prevent disease, with the ultimate goal of improving population health. Data analytics can also facilitate monitoring of medical devices and their performances, thus creating an opportunity for intervention when necessary (HealthIT.gov 2015). Much emphasis is placed on learning health systems which are somewhat closed loop environments that connect the healthcare delivery system to the community it serves through the flow of electronic information (HealthIT.gov 2015). Data analytics can certainly support such learning environments by relating and analyzing data from different sources (from patients, providers, and health insurance companies). A substantive case is made for data analytics from a perspective of supporting wearable devices and information that is collected from patients by using mobile devices, as in the case of patient portals or other disease management applications (Gay and Leijdekkers 2015). Big Data analytics can also contribute to lowering healthcare expenditures. According to McKinsey Global Institution (McKinsey and Company 2011), it is estimated that Big Data Analytics has a potential value of $300 billion to US healthcare.

## 21.3 Big Data Analytics

According to Merriam-Webster dictionary, analytics pertains to analysis or separating something into smaller components. Big Data is the term used to describe a large collection of digital data, initially characterized by volume, variety, and velocity that require advanced technologies and techniques for capture, management, and analysis (Institute for Health Technology Transformation 2013). Data collected in healthcare includes notes and records from the patient monitoring systems, imaging systems, medication orders and medication administration, clinician reports, quality reports, billing and financial reports, patient-originated records, and more, all of which constitute an incredible volume of data. Variety of data is exhibited by the different formats in which it is collected and stored, such as wired or wireless health monitors, image, video, numbers, templates, pick lists or dropdown menus, scanned handwritten notes, and text found in doctors' notes or e-mails between patients and providers. Big Data in health care can be categorized as structured and unstructured (Sayles 2013). Structured data represents numbers, text, or other values that are typically stored in relational databases. Given the variety in the structure of data, some call this category multi-structured data (Arthur 2013). Unstructured data typically include free text, videos, images, or audio data. Given the varied degrees of structure, the term semi-structured data has also emerged. For example, a consult report recorded in a template by the doctor is text and appears to be unstructured; however, it has an organizational structure that's provided by the template, which gives it some structure and makes it easier to organize and analyze. Velocity is a data characteristic that reflects the speed of data creation, data processing, storage, and analysis.

In the last few years, more "Vs" have been introduced to fully describe Big Data: variability, veracity, visualization, and value. Veracity represents data accuracy and credibility. For various committees focused on data standards and data quality, veracity is definitely a desired characteristic but it is not an inherited characteristic of Big Data (HealthIT.gov 2014). Variability reflects potential changing of data meaning. Think about patients with same diagnosis that may have been coded differently because of medical coding pitfalls. This may affect the accuracy of reporting, billing, and any decisions that depend on such medical codes. Data visualization refers to the way data is presented in order to make it easy to understand. Consider the interactive disease maps that have become available at the Center for Disease Control website or the hospital compare databases that allow users to easily select a certain number of variables and compare hospitals of choice. Data value refers to the potential of certain data to impact quality and cost of healthcare.

In order to understand the Big Data journey from collection to end reports, Raghupathi and Raghupathi (Raghupathi and Raghupathi 2014) developed a conceptual framework that comprises of data moving from its original source to a transformation-type application, from its transformed format into a Big Data platform, and from there into a specific analytics application. Let's look at this process in greater detail.

Big Data is generated from multiple sources. From a perspective of an organization, those resources can be internal or external. Internal resources include data from the electronic health records, computerized physician order entry, and other systems within the organization. External sources include data from external lab services or pharmacies, health insurance companies, or government sources. As already established, sources of data and types of data are different and they also come from different applications, such as a transaction processing system or picture archiving and communication system. Data could also come from various locations, as in the case of clinics or other hospitals in the network that transfer among each-other data. Another illustration of external data are state cancer or immunization registries that receive data from multiple providers/locations throughout the state.

The data generated from these different sources is considered raw data and it typically goes through a transformation or pooling process that can also be accomplished in different ways (Raghupathi and Raghupathi 2014). One popular method is a data warehousing, which aggregates data and makes it ready for processing. A second method is pooling the information in tables of the CSV format. A third method is to extract, transform, and load (ETL) the data by using ETL software which is a program that extracts the data from various sources, transforms it to fit the needs, and loads it into a data store or data mart where the data is operational. The fourth method is to transfer the data into a middleware, which is an application that serves as a bridge between the original (initial) database and the analytics applications. Once the data is transformed into one of these applications, it is ready to be used as input for platforms or tools that enable data analysis.

There are various platforms and tools that enable data analytics. One of the most popular platforms that is considered an indispensable environment for Big Data Analytics is Hadoop (Rajeshwari 2015). The Hadoop Distributed File System, one of the Apache™ Hadoop projects provides high-throughput access to data by organizing and dividing the data into smaller clusters and distributing it to various servers (Hadoop 2016). Various servers manipulate smaller chunks of the data and seek to solve different parts of a larger problem. Once that process is completed, Hadoop has the potential to bring those parts together and produce an integrated final result. Despite its potential in data analytics, especially given that it is open source and it can be used in analyzing unstructured data, Hadoop is not easy to install, configure and use. Healthcare organizations have yet to fully embrace Hadoop. MapReduce is a tool that follows a programmed logic to filter and organize data into smaller chunks; thus addressing scalability of data and making data analysis process more efficient. There are other tools (most of which work in conjunction with Hadoop) that provide a platform for Structured Query Language (SQL), such as Hive; or provide a platform for NoSQL, such as Such as HBase, Cassandra, and Lucerne (Raghupathi and Raghupathi 2014). Other platforms, such as PIG and PIG Latin can be used to assimilate structured or unstructured data. One of the trends in healthcare is the adaption of NoSQL technologies which are utilized to analyze unstructured data (Tableau 2016). Healthcare data is text-heavy and despite the implementation of electronic health records, a good portion of the data is unstructured. In addition, patient generated data on mobile devices add to the complexity of data structure. NoSQL technologies create platforms for queries, data mining and other reports in order to take advantage of the information rich unstructured data.

Finally, the analytics platform makes it possible to complete the data analytics process by running queries, reports, data mining, or online analytical processing (OLAP). Queries are applications that allow creation of report templates. Data mining is the process of searching through large amounts of data in order to look for patterns or relationships (Sayles 2013). The search logic is programmed based on the type of data. OLAP refers to processing of the data during its manipulation for analysis (OLAP in the Data Warehouse 2001).

## 21.4 Mobile Analytics

Growth of mobile technologies and use of mobile apps have increased the amount of mobile generated health data. Mobile health market is projected to continue to grow. A Berks Insight report found that that 7.1 million patients were remotely monitored in 2016 and this number is predicted to reach 50 million in 2021 (Berks Insight 2017). Approximately 1.7 billion smartphone users will download at least one health app by 2018 (Petersen et al. 2015). The number of applications available for download provided by Apple or Android is more than 500,000 for each and some of the apps have data analytic capabilities (Chen et al. 2011). Most mobile devices have some kind of embedded mobile analytics tool. The most popular one is the mobile-sensing app, which comes with sensors such as accelerometer, digital compass, gyroscope, GPS, microphone, and camera (Lane et al. 2010). Apps like Pay Your Selfie, which pay consumer to complete selfie tasks, collect much information on consumer's activities, preferences, people, places, and products they like and use data analytics to understand their lifestyle trends and preferences. In addition, many businesses have created mobile apps that provide a better browsing, navigation or entertainment experience for the user. Insight from data analytics can result in increased traffic, gaming or even business revenue. Mobile health apps are not much different from the other types of apps. They offer a convenient way to collect and transmit data. Further, they can support analytics in order to monitor patients, intervene for timely care, prevent, or manage disease.

## 21.5 Future Considerations

Advancements in data analytics are helping improve patient care and patient care operations; however, there are still untapped opportunities. More work lies ahead, especially in regard to capitalizing in some types of data, such as medical images, medical signals, standardization, user-friendliness of analytics platforms, and data security. Images generated by X-rays, magnetic resonance imaging, ultrasounds, fluoroscopy, computed tomography and other imaging techniques comprise a large amount of health data. Analytics of such data can make a significant impact in patients' diagnoses, treatment and prognosis (Belle et al. 2015). The widespread of Picture Archiving and Communication Systems (PACS) among hospitals and clinics enhances the opportunity to capture, store and share imaging data. The high volume and larger storage space required for this type of data is challenging but decreasing cost of storage space should help address those concerns. Current technologies such TeraRecon iNtuition which enables federated queries and reviews across imaging archives open new horizons for imaging analytics and encapsulating the wide range of data.

Another untapped opportunity is analytics pertaining to signal processing. Patient monitors are part of every hospital and many clinics (Belle et al. 2015). Wireless monitoring devices, patient-generated health data from mobile devices, and wearable sensors are also expanding and could be connected to a signaling process. Currently, the alarm triggering process is mostly driven by system's reliance on a single source of information, such as patient's systolic and diastolic blood pressure readings. Given the specific clinical condition of each patient, those triggers are often not a good indication of the situation. This leads to either arbitrary alteration of the alarm system or "alarm fatigue", none of which supports patient safety. Analytic tools could make it possible to capture relevant data from other clinical systems, integrate and correlate them for a more accurate assessment and a more meaningful alert system.

For many years, healthcare organizations have worked on the implementation of data standards such as HL7, LOINC, RxNorm, DICOM and others. Such standards certainly facilitate data analytics. As the data pool has expanded to include mobile-generated health data, standards become questionable. Current regulation and standardization of mobile data is not at the same levels as those generated by non-mobile systems. Many of the analytics platforms have the capability to work with standardized or non-standardized data, however, this is another consideration in the discussion of data analytics.

Current data analytics tools available are complex and not easy to use. Simplification and improvement in terms of installation and usage could make a big difference for healthcare organizations. The manpower to work with the existing systems is limited and in high demand, which makes it difficult for some of the providers to engage in data analytics. In addition, the types of reports produced are not always user-friendly and may not contribute to better decision-making or process improvements on a timely fashion. Visualization of the results produced by queries, reports, OLAP, or data mining applications need to be intuitive and easy to understand. Only then, would Big Data Analytics become an ordinary part of operations.

Last but not least, one of the most challenging battles has been the need to use healthcare data and the ongoing public cry for privacy. Use of data analytics can certainly benefit individual and public health and enhance the quality of healthcare. On the other side, increase in data volume, combination of data sets, the possibility of individual re-identification after de-identification, and lack of thorough HIPAA regulations (some of the health-generated data is not covered by HIPAA) present serious privacy and security issues (HealthIT.gov 2015). Maintaining the proper balance is challenging but it is important to the public. As current data analytics platforms and tools evolve and new ones emerge, it is important to comply with federal regulations and enhance security features.

## 21.6 Summary

Increased volume and variety of data, as well as development of new technologies for data analysis have moved data analytics at a different level. The multitude of Big Data Analytics platforms, tools and applications available make it possible to search, cluster, identify trends and patterns and create different types of reports for structured and unstructured data. Results from data analytics can be used to improve health and healthcare for individuals and communities, as well as lower healthcare costs. Health data generated from mobile devices has the potential for great contribution to the existing data pools. While much progress has been made, there is still room for advancement towards better utilization of imaging and signal processing data, data standardization, and simplification of data analytics tools. In addition, data analytics tools must provide a secure environment for data management and analysis.

## References

Arthur L. What is big data? Forbes. 2013. http://www.forbes.com/sites/lisaarthur/2013/08/15/what-is-big-data/#2cc83a933487. Accessed 4 Nov 2018.

Belle A, Thiagarajan R, Soroushmehr SR, Navidi F, Beard DA, Najarian K. Big data analytics in healthcare. Biomed Res Int. 2015;2015:370194. https://doi.org/10.1155/2015/370194.

Berks Insight. mHealth and home monitoring. M2M research series. 2017. http://www.berginsight.com/ReportPDF/ProductSheet/bi-mhealth8-ps.pdf. Accessed 4 Nov 2018.

Chen H, Chiang RHL, Storey VC. Business intelligence and analytics: from big data to big impact. MIS Q. 2011;36(4):1165–88.

Gay V, Leijdekkers P. Bringing health and fitness data together for connected health care: mobile apps as enablers of interoperability. J Med Internet Res. 2015;17(11):e260. https://doi.org/10.2196/jmir.5094.

Hadoop. Welcome to Apache™ Hadoop. 2016. http://hadoop.apache.org/. Accessed 4 Nov 2018.

HealthIT.gov. Clinical quality. 2014. https://www.healthit.gov/facas/health-it-standards-committee/hitsc-workgroups/clinical-quality. Accessed 6 Feb 2018.

HealthIT.gov. Health big data recommendations: HITPC Privacy and Security Workgroups. 2015. https://www.healthit.gov/sites/faca/files/HITPC_Health_Big_Data_Report_FINAL.pdf. Accessed 6 Feb 2018.

Institute for Health Technology Transformation. Transforming health care through big data strategies for leveraging big data in the health care industry. 2013. http://ihealthtran.com/wordpress/2013/03/iht%C2%B2-releases-big-data-research-report-download-today/. Accessed 6 Feb 2018.

Lane ND, Miluzzo E, Lu H, Peebles D, Choudhury T, Campbell AT. A survey of mobile phone sensing. IEEE Commun Mag. 2010;48(9):140–50. https://www.cs.cornell.edu/~tanzeem/pubs/mobile_phone_survey.pdf Accessed 6 Feb 2018.

McKinsey & Company. Big data: the next frontier for innovation, competition, and productivity. 2011. http://www.mckinsey.com/business-functions/business-technology/our-insights/big-data-the-next-frontier-for-innovation. Accessed 6 Feb 2018.

OLAP in the Data Warehouse. Data warehousing fundamentals. Upper Saddle River: Prentice Hall; 2001. p. 343–75.

Petersen C, Adams SA, DeMuro PR. mHealth: don't forget all the stakeholders in the business case. Medicine 2.0. 2015;4(2):e4. https://doi.org/10.2196/med20.4349.

Raghupathi W, Raghupathi V. Big data analytics in healthcare: promise and potential. Health Inf Sci Syst. 2014;2(1):3. https://doi.org/10.1186/2047-2501-2-3.

Rajeshwari D. State of the art of big data analytics: a survey. Int J Comput Appl. 2015;120(22):39–46.

Sayles NB. Health information management technology. An applied approach. 4th ed. Chicago: AHIMA Press; 2013.

Tableau. Top 8 trends for 2016: big data. 2016. http://www.tableau.com/sites/default/files/media/top8bigdatatrends2016_final_2.pdf?ref=lp&signin=0312a8bdf6c23b6d092842c3c1afbdfd. Accessed 6 Feb 2018.

White House Big Data Report. Big date: seizing opportunities, preserving values. 2014. www.whitehouse.gov/sites/default/files/docs/big_data_privacy_report_may_1_2014.pdf. Accessed 6 Feb 2018.

# Chapter 22
# Mobility and Cloud Computing

**Egondu R. Onyejekwe and Hung Ching**

**Abstract** The concept of mobile health (mHealth) portends the ability to move health information. There is a surge of participation by third parties who enable the mobility of health records. They posit, to a bigger or lesser extent, that they have applied the appropriate methods for either electronic health records management (say in iCloud settings) and or have the mobile devices to capture care data efficiently. The paradigm shift to a Cloud computing environment portends several opportunities for healthcare service providers. There are opportunities to provide healthcare services, not only in different scenarios, but also in a diversity of effective and simple way. Furthermore, there are the advantages of *scalability* and *mobility* that a Cloud-based healthcare services environment system can offer.

**Keywords** Mobility · Cloud · iCloud · mHealth · Computing · Scalability

## 22.1 Introduction

The concept of mobile health (mHealth) portends the ability to move health information. This is neither limited by spatial nor temporal constraints. To achieve mobility with electronic health records then means that, they must be stored where they can easily be accessed by those who need them, especially the providers and the patients. Furthermore, there should be devices that can not only extract the necessary information at the point of need, but those that would also transport the health information as needed. For storage purposes, most people would look at Cloud computing and even flash drives. While for accessing and porting health

E. R. Onyejekwe (✉)
Public Health, Health Administration, College of Health Sciences, Walden University, Minneapolis, MN, USA
e-mail: Egondu.onyejekwe@mail.waldenu.edu

H. Ching
Department of Medical Physics, Memorial Sloan Kettering Cancer Center (National Institutes of Health Funded), New York, NY, USA
e-mail: chingh@mskcc.org

© Springer Nature Switzerland AG 2019
E. R. Onyejekwe et al. (eds.), *Portable Health Records in a Mobile Society*,
Health Informatics, https://doi.org/10.1007/978-3-030-19937-1_22

information, most would look at mobile devices. To enable mobile devices to fulfill these roles, they must be fortified with mobile apps. In today's world, the industry of mobile devices and care applications is growing at an alarming rate. But is such a growth commensurate with efficiently and securely porting or moving patient's data?

This chapter discusses the role of the "third party" providers of both the storage capabilities and the provision of care apps for mobile devices. Also, discussed are the users, including providers such as physicians, and clientele, such as the patients. The patients' privacy and security issues are addressed as well. While there are laws for regulations in these areas, they do have some short comings. The chapter concludes with some recommendations.

## 22.2 Role of Third parties

There is a surge of participation by third parties who enable the mobility of health records. They posit, to a bigger or lesser extent, that they have applied the appropriate methods for either electronic health records management (say in iCloud settings) and or have the mobile devices to capture care data efficiently. But for most provisions, the third party designers may not account for either patient security and or privacy of patient care data. Consequently, insecure applications may be released in the marketplace. In advanced countries such as the United states and the European Union, there are existing laws that regulate mHealth, but the creation of the mHealth applications may not satisfy these laws and regulations. This chapter serves as a brief guide for developers and designers of apps concerning issues relating to security and privacy standards and certifications.

## 22.3 Cloud Computing Paradigm

### 22.3.1 CureMD's EHR

mHealth can benefit from the features and functionalities that the Cloud Computing paradigm offers. CureMD uses the iCloud developed by Avalon for EHR management. CureMD consists of web-based EHRs that connect physicians to hospitals, labs, pharmacies, insurance companies, radiology, patients, payors, etc. with a simple login. Avalon provides this for a variety of practices, that range from solo to large multi-specialty, and other practices. From their perspective, CureMD is convenient for physicians who see a multiple number of patients daily. Those who use CureMD can access the patient's record by simple clicks. (They do not have to worry about the expenses of both installations and maintenance of hardware and software.) Essentially, Avalon states that CureMD delivers advanced features in the Cloud, at a fraction of the costs previously available in systems of such performance.

Basically, the physicians can log in from anywhere—as long as they have a device (a PC or other) and internet connectivity. Once logged in, they can create *Provider Notes* in three simple steps. The first step includes going to that day's (today's) patients and selecting a patient and creating the notes by choosing the relevant template; they choose between simply importing previous notes or auto-populating from previous visits with histories, allergies, reviewer system, etc. Next, they can do quick searches or set up the list with the different components. Where no lists are available, they simply have to ask their account manager to create lists on the fly. Thereafter, they can add physical exams, diagnoses, procedures as well as enter their plans. Thirdly, they can sign a superbill, so billers can submit bills electronically—tracked by CureMD—to see how many bills are paid or denied. These simple set of activities would now have created the provider notes and would also have submitted the superbill. Avalon surmises that CureMD, therefore, allows physicians to focus on their real passion—the practice of medicine!

Avalon, therefore, invites physicians to transform their practice(s) with CureMD which is a Cloud-based Electronic Health Record management system. They claim it has a powerful knowledgebase, that it is built for "usability, performance and reliability" (SelectHub 2018).

Their version 10g also includes *an iPad app* which according to Avalon "works delightfully…is intuitive, simple and enjoyable" (CureMD 2018a). So a physician now has bundled in one app, complete access to their appointments, clinical reviews, patient notes, medical histories, and document manager. Finally, they claim that with CureMD, practitioners do not have to spend thousands of dollars on set up, maintenance, and licensing fees. Neither do they have to pay for workflow-driven customization or dedicated support.

CureMD can bring about improved care by instantly connecting with "patients, payers, labs, hospitals, and other stakeholders" (CureMD 2018b). "CureMD EHR is ONC 2015 Edition certified and is MIPS, Meaningful Use Stage 2, and Meaningful Use Stage 3 ready" (CureMD 2018b). The EHR also has "an ICD-10 guarantee enabling you to stay ahead of the industry" (CureMD 2018b).

Other features of CureMD include: electronic prescribing; electronic labs; workflow automation; interoperability; and EHR for iPad and iPhone. Each of these is briefly described below.

## 22.3.2 Electronic Prescribing

Avalon also provides an electronic prescribing (ePrescribing) service which connects prescribers with over 40,000 pharmacies across the United States. Among the ePrescribing features are:

- Up-to-date drug knowledgebase
- Medication reconciliation
- Complete medication history from pharmacies

- Mail order and retail pharmacies
- Age and weight-based dose adjustment
- Process refill requests through patient portal and pharmacies
- Controlled substance e-prescribing
- Real time prescription eligibility and formulary (MediPro 2018)

Avalon claims advanced safety features that "ensure utmost quality and reliability" (CureMD 2018a). These are mediated by "providing access to prescription benefits, prescription history, formulary, eligibility, adverse reactions (drug-drug; drug-allergy, drug-diagnosis) and recommended dosages" (MediPro 2018).

### 22.3.3  Electronic Labs

Avalon also provides an advanced lab interfacing technology that allows practitioners to connect to all their favorite labs for sending orders and receiving results electronically from most of those labs. Among the features are those that will allow the physician to deliver safer and more reliable care. Among those are: trending and task assignment; comparing current and past results; alerts for abnormal results; and advanced reporting.

### 22.3.4  Workflow Automation

Avalon provides an integrated workflow which it claims also adapts to each unique preference(s) and practice style(s). Physicians can thus personalize their operation because the system mirrors the existing processes and revitalizes them with powerful automation and collaboration tools. Features such as "KPI dashboards, enterprise scheduling, intelligent billing, data mining reports and EHR," (CureMD 2019) and all other aspects of the system are integrated. They are also customizable to increase learning, adoption, and service throughput. Invariably, Avalon posits that these drive:

- Improve productivity
- Accelerate revenue cycle
- Decrease cost and risk
- Optimize collaboration
- Enhance service quality
- Ensure compliance
- Engage patients (MediPro 2019)

### 22.3.5 Interoperability

Avalon claims that the design of CurcMD allows for interoperability. CureMD, therefore, enables the seamless exchange of information between the different stakeholders listed below:

- Health information exchanges
- Pharmacies
- Payers
- Radiology/imaging services
- Hospital networks
- Referring providers
- Cancer registries
- Patients
- Syndromic surveillance agencies
- Immunization registries
- Specialty registries
- Electronic devices
- DICOM compliant imaging equipment (MediPro 2018)

Avalon surmises that it has created a platform for the future. This includes everything a physician needs and wherever they need it! It claims a fully-featured platform that will support a practitioner's entire practice life cycle. The platform is where a physician can chart on their iPad/iPhone "while engaging with patients at the same time" (CureMD 2018c). This is synchronized with CureMD EHR. It therefore, allows the practitioner to practice on the go as they can do any of the following:

- Document images
- Verify eligibility
- Create demographics
- Create new patient records
- Dictate through Siri
- ePrescribe
- Schedule patients
- Review clinical records (Skyrose 2018)

The paper concludes that Avalon has provided an innovative architecture built for doctors on the go who are freed from documenting and other hassles. Avalon claims that their design addresses is the "providers' convenience and ease of use" (Skyrose 2018).

### 22.3.6   All-New Design

Avalon's new design includes a "more capable system" (Skyrose 2018) the new system that is also integrated with the provider's CureMD EHR. Even where the modules look different, Avalon claims that the way any provider conducts their business will still feel instantly familiar.

### 22.3.7   iPad and iPhone EHR App

Avalon has added iPad and iPhone app that would promote a practitioner's usability, intuitiveness, and enjoyability experiences. Avalon makes it possible for practitioners to do the following:

- See daily appointments
- View clinical summaries
- Create patient notes
- Collect charges
- Manage documents
- Verify insurance eligibility (CureMD 2018a)

### 22.3.8   Privacy and Security Issues: Consequences of Cloud Computing Paradigm

The paradigm shift to a Cloud computing environment portends several opportunities for healthcare service providers. There are opportunities to provide healthcare services, not only in different scenarios, but also in a diversity of effective and simple ways. Furthermore, there are the advantages of *scalability* and *mobility* that a Cloud-based healthcare services environment system can offer. A major advantage of Cloud computing is the ability to share patient records with a diversity of providers: clinics, hospitals, doctors' and specialty offices, other healthcare providers and centers which include laboratories, pharmacies, radiology, etc. A critical point here is the integration of all the information in the EHR. This enables the different segments of the healthcare and medical staff to perform their jobs.

However, despite all the claims for the value of Cloud computing by Avalon and other third parties, there are a variety of security and privacy risks associated with moving sensitive patient health records to the Cloud. These security and privacy issues should be of concern to both the healthcare provider and to the Cloud Services provider. For starters, this is a different computing paradigm. So, there should be a solid relationship built on trust with the Cloud service provider to ensure a transparent process with the healthcare provider. Cloud service providers must assure that all security mechanisms are in place in order to avoid unauthorized access and data breaches. Also, the provider(s) must deploy transmission and network secure protocols to avoid external attacks to the data whether in transit or while being ported. Additionally, and to further ensure the privacy of sensitive patient health data and information, the Cloud service providers must also deploy authentication systems.

Patients should be integral parts of the loop and must be kept informed and updated about the management of their health data. In any event, whenever a Cloud-based paradigm is deployed for health records, the privacy and security of sensitive health data that has been migrated to the Cloud must be addressed as these constitute the main barriers for any given Cloud computing paradigm.

Among the generic security issues that must be addressed by both Cloud service providers and their healthcare providers are: role-based access, network security mechanisms, data encryption, digital signatures, and access monitoring.

## 22.4  The Laws

Also, and in order to guarantee the safety of the health information and comply with privacy policies, the U.S requires the Cloud service provider to be compliant with various certifications and third party requirements (Chaparro et al. 2017). Among those are: SAS70 Type II, PCI DSS Level 1, ISO 27001, and the US Federal Information Security Management Act (FISMA) (Chaparro et al. 2017).

Overall, governments everywhere must require that Cloud service providers fulfill the necessary privacy requirements to ensure the privacy of patient data (Rodrigues et al. 2013). One of the benefits of deploying a legal framework is to assure a secure environment. Privacy policies, especially in advanced countries, are legislated both to regulate and safeguard the privacy of patient health records.

In the US, for example, the Health Insurance Portability and Accountability Act (HIPAA), discussed elsewhere in this book, was created to regulate the privacy and security of US patient data. These policies depend on each country. Beyond what countries can do, there are other privacy and security terminologies included in standards, such as Health Level 7 (HL7) that target electronic health records (EHRs). For a patient then, a *secure health Cloud* environment includes the combination of the specific country's regulations with the widely accepted standards and Cloud policies and security mechanisms (Rodrigues et al. 2013).

## 22.5  Access Devices

Mobile phone health apps may now seem to be ubiquitous, yet much remains unknown with regard to their usage. Usually these apps become popular through word of mouth and advertising. Information is limited with regard to important metrics, including the percentage of the population that uses health apps, reasons for adoption/nonadoption, and reasons for noncontinuance of use.

Mobile health inevitably includes the mobile applications (apps) to the health-care delivery system in order to improve lives. These would involve the use of mobile apps (software) on smartphones for different aspects of healthcare delivery. Mobile apps can be applied to a variety of health-related issues. Discussed here are the mobile health applications to health-seeking behavior and to a chronic disease (such as diabetes) self-management. Chronic diseases, such as high blood pressure, hypertension, diabetes and certain types of cancers, among others, can be made manageable with the support of mobile apps.

One such application on smartphones can also help people with diabetes to control their fitness and health. Brzan, Rotman, Pajnkihar, and Klanjsek conducted a systematic review of access devises for mobile and wireless applications regarding the control and self-management of diabetes (Brzan et al. 2016). Their focus was on the review of free apps using the most common and popular mobile devices app stores that include Google Play (Android), App Store (iOS), and Windows Phone Store, from November to December 2015. The review specifically addressed these freely available mobile apps for self-management of diabetes. The intent was to assess the promotion of diabetes self-management as defined by Goyal and Cafazzo, following the Preferred Reporting Items for Systematic Reviews and Meta-Analyses (PRISMA) (Brzan et al. 2016). The self-management attributes of Goyal and Cafazzo include: monitoring blood glucose level and medication; nutrition; physical exercise; and body weight (Brzan et al. 2016).

Brzan et al.'s 2016 study is entitled Mobile Applications for Control and Self Management of Diabetes: A Systematic Review. They included three independent experts in the field of healthcare-related mobile apps for the assessment for eligibility and testing phase. A total of 65 apps that included 21 from Google Play Store, 31 from App Store and 13 from Windows Phone Store were tested and evaluated (Brzan et al. 2016).

According to Brzan et al., of the 65 apps, "fifty-six of these apps did not meet even minimal requirements or did not work properly" (Brzan et al. 2016). Therefore, from the wide range of applications that claim relevance to the self-management of diabetes, their results showed that only nine (5 from Google Play Store, 3 from App Store, and 1 from Windows Phone Store) out of 65 reviewed mobile apps, had such relevance (Brzan et al. 2016). Those nine apps were assessed to be versatile and useful for successful self-management of diabetes based on selection criteria of the study. Of course, it is possible to differentiate the levels of inclusion of features based on selection criteria of any set of selected mobile applications.

The pertinent aspects of this study's results are that the findings can be used as discriminants for what content(s) app developers need to include. In addition, certain specific recommendations, such as more features, for mobile apps for self-management of diabetes are in order. These will likely increase the number of long-term users, and therefore, has the potential to influence better self-management of diabetes and other chronic diseases.

## 22.6 Users and Uses

Use of mobile health applications for health-seeking behavior among US adults, was a study by Bhuyan, Lu, Chandak, Kim, Wyant, and Bhatt, that exploresd the use of mobile health applications (mHealth apps) on smartphones or tablets for health-seeking behavior among US adults (Bhuyan et al. 2016). They obtained secondary data from cycle 4 of the 4th edition of the Health Information National Trends Survey (HINTS 4) (Bhuyan et al. 2016). They applied weighted multivariate logistic regression models as predictors of four variables. The variables were classified as:

1. having mHealth apps
2. usefulness of mHealth apps in achieving health behavior goals
3. helpfulness in medical care decision-making
4. asking a physician new questions or seeking a second opinion (Bhuyan et al. 2016)

They used the Andersen Model of health services utilization to group the independent variables of interest under three major factors: predisposing factors, enabling factors and the need factors (Bhuyan et al. 2016). The *predisposing* factors included age, gender, race, ethnicity, and marital status. The *enabling* factors included education, employment, income, regular provider, health insurance, and rural/urban location of residence. The *need* factors included general health, confidence in their ability to take care of health, body mass Index, smoking status, and number of comorbidities (Bhuyan et al. 2016).

Their findings form the national sample of adults who had smartphones or tablets, included 36% who had mHealth apps on their devices. Among these 36% with apps, 60% reported the usefulness of mHealth apps in achieving health behavior goals, 35% reported their helpfulness for medical care decision-making, and 38% reported their usefulness in asking their physicians new questions or seeking a second opinion (Bhuyan et al. 2016).

The analyses of the multivariate models showed that respondents were more likely to have mHealth apps if they had more education, health insurance, were confident in their ability to take good care of themselves, or had comorbidities. Alternatively, the multivariate models analyses, showed that the respondents were less likely to have the mHealth apps if they were older, had higher income, or lived in rural areas (Bhuyan et al. 2016).

Regarding the usefulness of mHealth apps, the older respondents with higher income were less likely to report their usefulness in achieving health behavior goals. However, older, African Americans, who had confidence in their ability to take care of their health were more likely to respond that the mHealth apps were helpful in making a medical care decision and asking their physicians new questions or for a second opinion (Bhuyan et al. 2016).

The authors concluded that while mHealth apps may potentially reduce the burden on primary care, reduce costs, and improve the quality of care, practitioners and researchers should be mindful of the several personal-level factors that were associated with having mHealth apps as well as their perceived helpfulness among their users, ultimately *"indicating a multidimensional digital divide in the population of US adults"* (Bhuyan et al. 2016).

In their research work entitled, Health App Use Among US Mobile Phone Owners: A National Survey, Krebs and Duncan explored health app usage among mobile phone owners in the United States by conducting a cross-sectional survey of 1604 mobile phone users within the United States (Krebs and Duncan 2015). They applied a 36-item survey for assessing sociodemographic characteristics, history of and reasons for health app use/nonuse, perceived effectiveness of health apps, reasons for stopping use, and general health status (Krebs and Duncan 2015).

Their results indicated that a little over half (934/1604, 58.23%) of mobile phone users had downloaded a health-related mobile app. The most common categories of health apps used (at least used on a daily basis) were fitness and nutrition. The common reasons for mobile phone users not downloading the apps were lack of interest, cost, and concern about apps collecting their data. In addition, their results indicated that the users who were more likely to use health apps tended to be younger, have higher incomes, be more educated, be Latino/Hispanic, and have a body mass index (BMI) in the obese range (all $P < 0.05$) (Krebs and Duncan 2015). Respondents also viewed cost as a significant concern, and a large proportion of them indicated that they would not pay anything for a health app. Additionally, those who had downloaded health apps, had trust in their accuracy and were of the view that data safety was quite high. Most of these respondents felt that the apps had improved their health. At the other extreme were half of the respondents (427/934, 45.7%) who had stopped using some health apps, primarily because of high data entry burden, loss of interest, and hidden costs (Krebs and Duncan 2015).

The authors opined that their findings "suggest that while many individuals use health apps, a substantial proportion of the population does not, and that even among those who use health apps, many stop using them" (Krebs and Duncan 2015). The researchers, therefore, recommended that app developers need to better address consumer concerns, "such as cost and high data entry burden, and that clinical trials are necessary to test the efficacy of health apps to broaden their appeal and adoption" (Krebs and Duncan 2015).

Additionally, when Weber, Adams, Bernstam, Bickel, Fox, Marsolo, et al. studied the effects of incomplete data, the results were quite revealing. Weber et al. in the study entitled, Biases introduced by filtering electronic health records for patients with "complete data," they evaluated how "simple heuristic checks" for data "completeness" would affect the number of patients in the resulting cohort (Weber et al. 2017). They also captured the potential biases the incomplete data would yield.

To check for the presence of demographics, laboratory tests, and other types of data, the researchers started with a set of 16 filters. This was followed by systematically applying all $2^{16}$ possible combinations of these filters to the EHR data for 12 million patients. The patients were drawn from seven (7) healthcare systems and a separate payor claims database of 7 million members (Weber et al. 2017).

The results revealed EHR data with considerable variability in data completeness across sites as well as high correlation between data types. For example, their results showed "the fraction of patients with diagnoses increased from 35.0% in all patients to 90.9% in those with at least 1 medication" (Weber et al. 2017). Additionally, they reported that an "unrelated claims dataset independently showed that most filters select members who are older and more likely female and can eliminate large portions of the population whose data are actually complete" (Weber et al. 2017). They cautioned, therefore, that investigators who design similar studies, should "balance their confidence in the completeness of the data with the effects of placing requirements on the data on the resulting patient cohort" (Weber et al. 2017).

## 22.7 Tailoring: To Specialty

There are several ways to address these problems, and tailoring to specialty, such as nursing is one way the Canadian Health Outcomes for Better Information and Care (C-HOBIC) project addressed it (Hannah et al. 2009). They did by introducing a systematic use of standardized clinical nursing terminology for patient assessments. C-HOBIC, implemented to date, in three Canadian provinces, comprises "an innovative model for large-scale capture of standardized nursing-sensitive clinical outcomes data within electronic health records (EHRs)" (Hannah et al. 2009).

This activity is supported by mapping nursing assessment and outcomes concepts to the International Classification for Nursing Practice (ICNP®) (Hannah et al. 2009). Serial data on a patient was compared across multiple time points. This allowed the C-HOBIC model to generate nursing-sensitive patient outcome reports. One major benefit of this C-HOBIC model is the information it provides to nurses, such as nursing information in either provincial databases or EHRs in the three Canadian provinces (Hannah et al. 2009). Such information promotes both the continuity of patient care across sectors of the healthcare systems in those provinces. Furthermore, it facilitates aggregation and analysis by administrators and policy makers, thus rendering information that is critical to planning for and evaluating patient care in nursing. Finally, the C-HOBIC model is attributed with the provision of "standardized, consistent, interoperable clinical information" (Hannah et al. 2009). The claim, of course, is that this reflects nursing practice "throughout the Canadian healthcare System," although the study was limited to three Canadian provinces (Hannah et al. 2009). Yet, it is a step forward.

## 22.8 Recommendations

As discussed above the Cloud Computing paradigm enables the enhancement of opportunities for electronic health records (EHR) systems and the extension of their functionalities including portability and mobility of patient records. Inherent, however, in the mobility of patient electronic health records are several risks attributable to the security and privacy of such records. Furthermore, there are the risks associated with the hosting of patients' Electronic Health Records (EHRs) on the servers of third party Cloud service providers. Some suggestions especially regarding the privacy and security issues that Cloud service providers should address in their platforms, have been articulated. Also addressed are the different roles required by provider and the Cloud Computing third parties to both protect the confidentiality of patient information, as well as facilitate the process of the mobility of such data.

A rehearsal of these include: role-based access, network security mechanisms, data encryption, digital signatures, and access monitoring. In addition, the Cloud service provider is required to be compliant with various certifications and third-party requirements. Among those are: SAS70 Type II; PCI DSS Level 1; ISO 27001; and the US Federal Information Security Management Act (FISMA)) (Chaparro et al. 2017). These are essential to guarantee the safety of the information and comply with privacy policies.

All these still make the promise of nationwide adoption of electronic health records (EHRs) very elusive. First, the EHRs still reside in pockets of the third party providers. The basic fundamentals remain the same as having the health records on folders, or on individual servers that belong to different entities. This silo mentality creates more problems than they are capable of solving. One problem is that the EHR data is not available for large-scale clinical research studies. A second problem is that the patient's EHR is fragmented since a given patient could be could be treated at multiple healthcare institutions, and each provider would only access what is available to them through their system. Consequently, the patient's EHR data from only a single site might not contain a composite and complete medical history for that patient. As a consequence, there is the potential for critical health events to be missing for that patient. All of problems increase inefficiencies in both the type and level of service provided and the attributed costs.

# References

Bhuyan SS, Lu N, Chandak A, Kim H, Wyant D, Bhatt J, et al. Use of mobile health applications for health-seeking behavior among US adults. J Med Syst. 2016;40(6):153. https://doi. org/10.1007/s10916-016-0492-7.

Brzan PP, Rotman E, Pajnkihar M, Klanjsek P. Mobile applications for control and self management of diabetes: a systematic review. J Med Syst. 2016;40(9):210. https://doi.org/10.1007/ s10916-016-0564-8.

Chaparro JD, Classen DC, Danforth M, Stockwell DC, Longhurst CA. National trends in safety performance of electronic health record systems in children's hospitals. J Am Med Inform Assoc. 2017;24(2):268–74. https://doi.org/10.1093/jamia/ocw134.

CureMD. Electronic health records (EHR). 2018a. http://www.curemd.com/ehr.asp. Accessed 1 Nov 2018.

CureMD. How electronic health records improve patient care? 2018b. https://www.curemd.com/ ehr-improve-patient-care.asp. Accessed 1 Nov 2018.

CureMD. Avalon. 2018c. https://www.curemd.com/avalon.asp. Accessed 1 Nov 2018.

CureMD. The smart cloud. 2019. https://www.curemd.com/EMR_Smart_Cloud.asp. Accessed 20 Jan 2019.

Hannah KJ, White PA, Nagle LM, Pringle DM. Standardizing nursing information in Canada for inclusion in electronic health records: C-HOBIC. J Am Med Inform Assoc. 2009;16(4):524–30. https://doi.org/10.1197/jamia.M2974.

Krebs P, Duncan DT. Health app use among US mobile phone owners: a national survey. JMIR Mhealth Uhealth. 2015;3:e101. https://doi.org/10.2196/mhealth.4924.

MediPro. EHR medical software programs. 2018. https://www.medipro.com/ehr-medical-software-programs/. Accessed 20 Jan 2019.

MediPro. Medical billing software 2019. 2019. https://www.medipro.com/blog/. Accessed 20 Jan 2019.

Rodrigues JJ, de la Torre I, Fernández G, López-Coronado M. Analysis of the security and privacy requirements of cloud-based electronic health records systems. J Med Internet Res. 2013;15(8):e186. https://doi.org/10.2196/jmir.2494.

SelectHub. CureMD EHR. 2018. https://selecthub.com/ehr-software/curemd-ehr/. Accessed 19 Jan 2019.

Skyrose. Full CureMD EHR software review-all you need to know about CureMD. 2018. https://skyose.com/full-curemd-ehr-software-review-all-you-need-to-know-about-curemd/. Accessed 21 Jan 2019.

Weber GM, Adams WG, Bernstam EV, Bickel JP, Fox KP, Marsolo K, et al. Biases introduced by filtering electronic health records for patients with "complete data". J Am Med Inform Assoc. 2017;24(6):1134–41. https://doi.org/10.1093/jamia/ocx071.

# Part VII
# Future Directions

# Chapter 23
# Foresight

Hung Ching

**Abstract** We must be cognizant of the need for the public to be properly informed through health education. Health literacy plays a very important role in adoption by the public. Associated with any big transformations in society, there will be resistance from people and institutions that are opposed to change. Surprisingly, adoption of totally integrated systems using EHRs in medical offices and hospitals has been slow. The current situation in the United States is still far from the plan that both former Presidents Bush and Obama had envisioned, but no other nation has moved as fast as the United States in adopting EHRs.

**Keywords** Foresight · EMR · EHR · PHR · HIT · Adoption · Health education · Health literacy

## 23.1 Introduction

As we progress with the advancement of changing and improved technology in portable healthcare, we must be cognizant of the need for the public to be properly informed through health education. Health literacy plays a significant role in adoption by the public. Associated with any big transformations in society, there will be resistance from people and institutions that are opposed to change. For example, if medical insurance companies see a reduction or even a threat to their profitability, they will definitely be resistant to change. In the same way, if companies that are heavily invested in electronic health records (EHRs) observe any decreased profits because of change, they will certainly do whatever they can to prevent any changes from happening.

H. Ching (✉)
Department of Medical Physics, Memorial Sloan Kettering Cancer Center (National Institutes of Health Funded), New York, NY, USA
e-mail: chingh@mskcc.org

## 23.2   What Is Health Literacy?

Health literary refers to the ability of people to obtain access to, comprehend, and utilize health information to cultivate and support positive health outcomes. These outcomes are usually related to beneficial health education activities. Good health literacy is exemplified by increased knowledge and comprehension of health determinants, improved mindset and attitude towards healthy behavior, and enhanced self-efficacy with respect to well-defined healthy activities (Nutbeam 2006).

## 23.3   Use of Health Literacy Terminology

Parker, Baker, Williams, and Nurss described health literacy as a person's ability to use his or her literacy competence to understand information concerning his or her health and medical information (Parker et al. 1995). Examples of the materials to be comprehended are items such as drug labels, instructions on how medication is to be taken, side effects from taking the medication, and things as simple as appointment cards for visiting medical offices. So by this very narrow definition of health literacy, the focus is on the patient's ability to adhere to the doctor's prescribed medical regimens (Nutbeam 2006).

## 23.4   Classifications of Literacy

The original way the words health literacy had been used was not only narrowly defined, but also was void of the more profound significance and intent of literacy for the general public. There are three types of literacy, and they differ by what it is that literacy allows the person to achieve.

### 23.4.1   Basic/Functional Literacy

Basic/functional literacy refers to the ability to function in an effective manner by reading and writing in everyday situations. This type of literacy is very similar to the narrow definition of health literacy noted above.

### 23.4.2   Communicative/Interactive Literacy

Communicative/interactive literacy refers to the ability to use more progressive cognitive and functional skills in conjunction with social skills in everyday situations.

It also refers to the ability to figure out meaning and extract information from various modes of communication. Moreover, this form of literacy allows a person to effectively utilize new information and apply them to dynamic situations.

### 23.4.3 Critical Literacy

Critical literacy refers to the ability to use more progressive cognitive and social skills to critically evaluate materials and information. It also refers to the ability to use newly analyzed information to gain greater control over different situations and life events (Nutbeam 2006).

## 23.5 Importance of Health Education

To illustrate the importance of health education, let's look at an event that actually occurred in a New York City dental office in the summer of 2018. The details revolving around what had transpired in this healthcare facility show us how crucial health education is, in the healthcare industry. The issue was concerning the potential dangers of scattered radiation at the dental office emanating from the dental X-ray machines.

One of the secretaries at the dental office wanted to make sure that she would be far away from the dental X-ray machine while the machine is producing X-rays. She had this concern for scattered X-rays as she was seven months pregnant. The pregnant secretary made sure that she exited the dental office and stood outside in front of the dental office while the X-rays were being taken. She was just very protective of her baby and did not want any kind of radiation to potentially harm her baby. Were her actions warranted? Had she been better educated about the radiation, would she have acted the way she had done during the exposure of X-rays at the office?

It turned out that the pregnant secretary was receiving more radiation dose from the sun and cosmic rays while she was standing outside in front of the dental office for 1 min. How can that be? It turns out that the radiation levels in New York City is about 2.0 mrem in one day for everybody due to background radiation. This number is less in some areas of this country and more in other areas. So in 1 h, the radiation dose received by standing outside is 2.0 mrem divided by 24 h, which is 0.0833 mrem. In 1 min, the radiation is 0.0833 mrem divided by 60 min which is 0.0014. Therefore, by staying outside for 2 min, the pregnant secretary received 0.0028 mrem of radiation.

It is ironic that had she stayed in the office for those 2 min, she would not have received the 0.0028 mrem of radiation from the sun and cosmic rays. She would have only received the radiation dose due to the scattered radiation from the dental X-ray machine, which only amounted to 0.0010 mrem at best. Believe it or not, she was actually receiving almost three times the radiation dose outside of the dental office. It is now slowly becoming apparent that she should have stayed put inside the dental office.

Health education, had she been exposed to some of relevant information for radiation, could have helped her to make a wiser choice for her and her baby. To illustrate this point even further, it is unfortunate that there is a chemical dose associated with the air pollutants from car exhausts as vehicles were passing by in the front of her dental office. So the pregnant secretary received an extra form of dose from pollutants such as nitrogen oxide and nitrogen dioxide without even realizing it. This real occurrence at a healthcare facility clearly illustrates how health education can allow people to make the right decisions for their own health and the health of others.

## 23.6   Need for EHRs and Technology

The weakness of the health information infrastructure became apparent and evident during the days of the Katrina disaster in Louisiana, which had occurred in 2005. Whether it is a natural disaster such as Hurricane Katrina or a man-made one such as a weapons of mass destruction attack, members of the general public need to have confidence that they can rely on the health information infrastructure to access information concerning their health. What the public does not need is one disaster leading to another disaster. Natural disasters and terrorist attacks are non-preventable to some extent, but a disaster in the health information infrastructure is certainly preventable. It is fortunate that catastrophic disasters do not occur too frequently, but this definitely does not reduce the need for a strong and reliable health information infrastructure to provide the necessary health information through the use of EHRs and technology (Tang et al. 2006).

## 23.7   Barriers Against Adoption

The Institute of Medicine has commented that the large-scale proliferation of EHRs could be critical to improving patient healthcare. EHRs can also possibly lower the financial costs of administering outpatient medical care. Despite general agreement on the usefulness and benefits of EHRs and health information technology (HIT), physicians have moved too slowly to adopt them. Doctors are the essential front-line users of EHRs and HIT, and there are signs of resistance to adoption from this group (Ajami and Bagheri-Tadi 2013).

## 23.8   HITECH Act of 2009

Due to the ratification of the 2009 Health Information Technology for Economic and Clinical Health (HITECH) Act, there has been a dramatic increase in the number of hospitals in the United States that started to use EHRs. It is hopeful that HIT would improve patients' quality of care. HIT tools empower patients to take control over their own health and health issues. Through technological advances, EHRs can bring healthcare providers closer to their patients by providing a centralized location for the purpose of communication and treatment. EHRs and HIT tools are breaking down barriers previously hindering patients from obtaining direct access to their own personal health records (Mackert et al. 2016).

## 23.9   EHRs, PHRs and EMRs

Personal health records (PHRs) have been around for decades. The most familiar form of a PHR is an immunization card that enumerates a person's immunization history. During the last 20 years or so, advancing technologies in medical care have spurred the growth of electronic medical records (EMRs). EMRs are just a narrower view of patients' EHRs with a focus on medical history rather than the overall health history. PHRs have also gone electronic, and the electronic version of the PHR is the HER (Bonander and Gates 2010).

## 23.10   Importance of Health Education in School

Education concerning health management should start early in an individual's life. Even starting in elementary school, health management techniques can be taught to students using very fundamental tools. After the groundwork is laid in primary school, additional reinforcement tools can be used at the primary school level as well as the secondary school level to teach students the benefits of cultivating an accurate and effective PHR. The health education in school occurs simultaneously while students themselves are experiencing teachable moments in life when they are sick or preoccupied with the health of friends and family members. When they become adults, they will have even more teachable moments to reflect upon and to learn from these personal health experiences.

Health education should continue on at the university/nursing school level, graduate school/medical school level, and even at the post-doctoral/residency level. School curricula should include didactic teaching and hands-on training concerning PHRs and EHRs. Courses and training should be focused on teaching providers techniques in educating patients about PHRs and EHRs and the importance of maintaining and managing their health records (Tang et al. 2006).

## 23.11   Medical Education

Medical education is transforming at a very swift pace, and this is due in part to the fast-changing and rapid advancement of health information technology in the healthcare industry. Medical and nursing students at the university and graduate levels have been incrementally utilizing EMR to help them administer care in the medical field. Carefully implemented EMRs can facilitate education to help future physicians and nurses to utilize evidence-based medicine in the medical treatment of patients (Tierney et al. 2013).

## 23.12 Adoption of EHRs

Surprisingly, adoption of totally integrated systems using EHRs in medical offices and hospitals has been slow. For those medical doctors that have been using EHRs in their healthcare setting, there has been immense satisfaction with the versatility of EHRs. There is also agreement that EHRs can help lower the risk of possible errors and better patient healthcare.

In 2008, *The New England Journal of Medicine* disclosed that 82% of EHR users experienced enhanced clinical decision-making, 82% of EHR users reported a decrease in medication errors, and 92% of EHR users experienced improved communication with other physicians and their patients (DesRoches et al. 2008). Even though there are clear advantages in using EHRs, it was not until the 2009 HITECH Act had been passed that there was a dramatic increase in the number of hospitals and medical offices that started to use EHRs (Palabindala et al. 2016).

There are many major obstacles to the implementation of EHR systems that must be resolved by leadership prior to hospital-wide or organization-wide adoption. Prior to transitioning to a fully integrated EHR system, hospitals and institutions must determine which of the medical and administrative employees should be dedicated to implementation project. The EHR vendor plays a crucial part in the entire transitioning process from the old charting system to the new EHR system. It is imperative that data integrity is not comprised during the seamless conversion, and a well-planned workflow design will lower the risk of errors.

If the transition phase is not conducted in proper manner, the chances of mistakes occurring will be elevated. Providers and the institution may be sued for malpractice and many other legal complications may arise. Because of potentially poor implementation, mistakes, misadministration, and even mortality may be elevated when a new EHR system is installed. What are the possible causes of these errors? Errors may be introduced into an EHR system by employees who are not familiar with the new system. System-wide crashes in a new EHR system may create adverse situations in patient care. These crashes may also prevent access to pertinent patient information.

If a new EHR implementation goes according to plan, then the conversion is simply just a temporary nuisance with temporary obstacles. If the implementation is delayed due to unforeseen circumstances, then clinical processes may be adversely affected, decision-making may be negatively influenced, and patient healthcare may potentially compromised. Moreover, the Health Insurance Portability and Accountability Act (HIPAA) unambiguously states that hospitals and institutions are completely responsible for their own EHR systems and all matters revolving around these systems (Palabindala et al. 2016). Hospital leadership must undertake the task of training all EHR users, and it must focus on continuing to upgrade the skills of these users. In this way, leadership can maintain competency among the users and user errors can be caught in time before a misadministration occurs (Palabindala et al. 2016).

EHRs can accumulate, for all practical purposes, an infinite number of flawlessly legible and immediately accessible records that include almost all aspects of care, regardless of when and where the care had been administered. Patients treated at early adopter hospitals such as Kaiser Permanente Northwest and the United States Department of Veterans Affairs, have already collected massive amounts of patient data and information in the EHRs (Sittig and Singh 2011). The enormous volume of information in a patient's EHR may mask some significant clinical findings even though access to the information is still very reliable. For example, a provider may miss an important aspect of the patient's EHR due to information overload. If this oversight somehow affects clinical decisions in the treatment process, the provider may be held accountable and be sued for negligence.

## 23.13  Cost of Adoption

The costs involved in purchasing and maintaining an EHR system are relatively high. For a private practice with five physicians, the cost incurred from the purchase of an EHR system is approximately $162,000. In addition, there is a $85,000 on average cost for the yearly maintenance of these systems (Millman 2014). Public spending related to EHRs has been fueled by the promising outlook of EHRs and the government's agenda to bring public health into the digital age. As of 2014, the federal government had already poured approximately $25 billion into the EHR sector (Millman 2014).

## 23.14  Current Status of EHR Adoption

In 2004, then-President Bush established a timeframe for the majority of Americans to obtain an EHR by 2014. In 2009, then-President-elect Obama reaffirmed just before his inauguration, his commitment to have all Americans use EHRs by 2014. The current situation in the United States is still far from the plan that both former Presidents Bush and Obama have envisioned, but no other nation has moved as fast as the United States in adopting EHRs (Millman 2014). The United States still has a way to go before reaching the goals established by the former two Presidents, but very good progress has been made so far. Progress will continue to be made in the area of EHR adoption.

## 23.15   Conclusion

In summary, physicians that have been using EHRs in their healthcare setting have experienced immense satisfaction with the versatility of EHRs. There is also agreement that EHRs can help lower the risk of possible errors and better patient healthcare. As we progress with the advancement of changing and improved technology in portable healthcare, we must realize the need for health education for the general public. Health literacy plays a significant role in adoption by the public of any new and innovative technology, and it refers to the ability of people to obtain access to, comprehend, and utilize health information to cultivate and support positive health outcomes. Basic/functional literacy refers to the ability to function in an effective manner by reading and writing in everyday situations. Communicative/interactive literacy refers to the ability to use more progressive cognitive and functional skills in conjunction with social skills in everyday situations. It also refers to the ability to figure out meaning and extract information from various modes of communication. Critical literacy refers to the ability to use more progressive cognitive and social skills to critically evaluate materials and information. This final classification of literacy also refers to the ability to use newly analyzed information to gain greater control over different situations and life events. Good health literacy is exemplified by increased knowledge and comprehension of health determinants, improved mindset and attitude towards healthy behavior, and enhanced self-efficacy with respect to well-defined healthy activities. Public spending related to EHRs has been fueled by the promising outlook of EHRs and the government's agenda to bring public health into the digital age.

## References

Ajami S, Bagheri-Tadi T. Barriers for adopting electronic health records (EHRs) by physicians. Acta Inform Med. 2013;21(2):129–34.

Bonander J, Gates S. Public health in an era of personal health records: opportunities for innovation and new partnerships. J Med Internet Res. 2010;12(3):e33.

DesRoches CM, Campbell EG, Rao SR, Donelan K, Ferris TG, Jha A. Electronic health records in ambulatory care – a national survey of physicians. N Engl J Med. 2008;359(1):50–60.

Mackert M, Mabry-Flynn A, Champlin S, Donovan EE, Pounders K. Health literacy and health information technology adoption: the potential for a new digital divide. J Med Internet Res. 2016;18(10):e264.

Millman J. Electronic health records were supposed to be everywhere this year. They're not – but it's okay. 2014. https://www.washingtonpost.com/news/wonk/wp/2014/08/07/electronic-health-records-were-supposed-to-be-everywhere-this-year-theyre-not-but-its-okay/?noredirect+on&utm_term=.c74e4a2cdb80. Accessed 15 July 2018.

Nutbeam D. Health literacy as a public health goal: a challenge for contemporary health education and communication strategies into the 21st century. Health Promot Int. 2006;15(3):259–67.

Palabindala V, Pamarthy A, Jonnalagadda NR. Adoption of electronic health records and barriers. J Community Hosp Intern Med Perspect. 2016;6(5):32643. https://doi.org/10.3402/jchimp.v6.32643.

Parker RM, Baker DW, Williams MV, Nurss JR. The test of functional health literacy in adults: a new instrument for measuring patients' literacy skills. J Gen Intern Med. 1995;10(10):537–41.

Sittig DF, Singh H. Legal, ethical, and financial dilemmas in electronic health record adoption and use. Pediatrics. 2011;127(4):1042–7.

Tang PC, Ash JS, Bates DW, Overhage JM, Sands DZ. Personal health records: definitions, benefits, and strategies for overcoming barriers to adoption. J Am Med Inform Assoc. 2006;13:121–6.

Tierney MJ, Pageler NM, Kahana M, Pantaleoni JL, Longhurst CA. Medical education in the electronic medical record (EMR) era: benefits, challenges, and future directions. Acad Med. 2013;88(6):748–52.

Debates on the application of informatics in the field of health care and its implications in the delivery of healthcare continue. A systematic literature review by
Roehrs, da Costa, da Rosa Righi, and de Oliveira (Roehrs et al. 2017) indicates the
following:

> Information and communication technology (ICT) has transformed the health care field
> worldwide. One of the main drivers of this change is the electronic health record (EHR).
> However, there are still open issues and challenges because the EHR usually reflects the
> partial view of a health care provider without the ability for patients to control or interact
> with their data. Furthermore, with the growth of mobile and ubiquitous computing, the
> number of records regarding personal health is increasing exponentially. This movement
> has been characterized as the Internet of Things (IoT), including the widespread develop-
> ment of wearable computing technology and assorted types of health-related sensors. This
> leads to the need for an integrated method of storing health-related data, defined as the
> personal health record which could be used by health care providers and patients. This
> approach could combine EHRs with data gathered from sensors or other wearable comput-
> ing devices. This unified view of patients' health could be shared with providers, who may
> not only use previous health-related records but also expand them with data resulting from
> their interactions. Another personal health record advantage is that patients can interact
> with their health data, making decisions that may positively affect their health.

Note that "PHR" in the above quote was replaced by "personal health record" in
order to avoid confusion with our definition of PHR. A more in-depth discussion of
the PHR is provided by Markle Foundation (Markle Foundation 2003):

> The Personal Health Record (PHR) is an Internet-based set of tools that allows people to
> access and coordinate their lifelong health information and make appropriate parts of it
> available to those who need it. PHRs offer an integrated and comprehensive view of health
> information, including information people generate themselves such as symptoms and
> medication use, information from doctors such as diagnoses and test results, and informa-
> tion from their pharmacies and insurance companies. Individuals access their PHRs via the
> Internet, using state-of-the-art security and privacy controls, at any time and from any loca-
> tion. Family members, doctors or school nurses can see portions of a PHR when necessary
> and emergency room staff can retrieve vital information from it in a crisis. People can use
> their PHR as a communications hub: to send email to doctors, transfer information to spe-
> cialists, receive test results and access online self-help tools. PHR connects each of us to the
> incredible potential of modern health care and gives us control over our own information.

In order to discuss the implementation issues of PHR, some background in the
development of the EMR is needed. For this reason EMRs are first discussed with
respect to implementation, costs, security, and interoperability. It should be noted that
it was a deliberate choice to exclude the implementation of PHR as a paper based
medical record system (though it is technically possible but highly inadvisable)
because portability would be severely restricted due to physical transmission issues for
a record of an ever increasing size (e.g. mail, fax, etc.). For example, storing genomic
information about a patient is a physical form would be costly and would eliminate the
use of that information by analytical processes required for interpretation.

   The natural progression towards PHR would be first to implement an EMR, support
interoperability with other EMRs used by other health care providers, standardize the
EMR storage system, and make it an open standard. When open standards for EMR
implementations are established there will be a common understanding of a format

that will open up the possibility for general open PHR implementations. Reference (Roehrs et al. 2017) provides an excellent overview of the state of development of PHR as of 2017, stating "Nonetheless, existing PHRs have limited intelligence and can only inform on a small set of users' health care needs" (see also (Luo et al. 2012)).

As noted above the focus is on implementing a universally accepted PHR using informatics tools, thus enabling the mobility of the information contained in the PHR by electronic means. Figure 24.1 shows an example of how mobile health is viewed today. The diagram can be extended by joining 'Health Records' and 'Mobile Devices' to 'Portable Health Records' noting that "Multiple EHRs for the same patient can coexist, but only one PHR would exist. The PHR can integrate data from many sources, ranging from devices connected to the patient to health data from EHRs stored in health care provider systems" (Silva et al. 2015).

For this reason a discussion of the costs and benefits of EMR implementation is first included. Only after a robust and tested EMR system should the progression to PHR occur.

Although EMR systems are not implemented by all practitioners they are now in use by a significant number of both individual practitioners, group practitioners, as well as hospitals and similar institutions. Furthermore there are many medical software vendors that offer EMR software.

Medical associations have also provided documents describing key features of EMRs. The description and regulations developed by The Canadian Medical Protective Association (Canadian Medical Protection Association 2018) is used as one of the bases for the discussion of EMR systems. The document provides general rules and best practices for EMRs, but "variations in practice are expected and may be appropriate". Key features of the document are sections "Patient consent and right to access" (p. 12) and "Data sharing and inter-physician arrangements" (p. 24). These features are also features that are key to implementing a PHR.

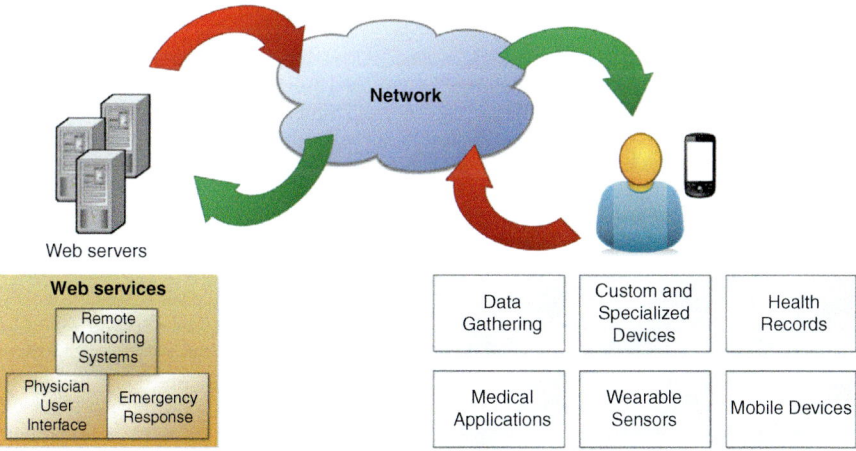

**Fig. 24.1** Illustration of a typical architecture of m-Health services (Silva et al. 2015). (Reproduced with permission of the copyright owner)

The Wikipedia article (Wikipedia n.d.) on EHR is another reference for EHR implementations. It has several sections on the cost of implementing EHR with comments such as "The steep price of EHR and provider uncertainty regarding the value they will derive from adoption, in the form of return on investment, has a significant influence on EHR adoption." The article further notes that: "In a project initiated by the Office of the National Coordinator for Health Information (ONC), surveyors found that hospital administrators and physicians who had adopted an EHR noted that any gains in efficiency were offset by reduced productivity as the technology was implemented, as well as the need to increase information technology staff to maintain the system" (Health IT. gov 2019).

In more detail (Wikipedia n.d.) discusses start-up costs, maintenance costs, training costs, software quality, usability deficiencies and lack of semantic interoperability. Clearly these costs are also incurred to an even greater extent for PHRs. For the individual physician the articles notes that "Office-based physicians in particular may see no benefit if they purchase such a product—and may even suffer financial harm. Even though the use of health IT could generate cost savings for the health system at large that might offset the EMR's cost, many physicians might not be able to reduce their office expenses or increase their revenue sufficiently to pay for it. For example, the use of health IT could reduce the number of duplicated diagnostic tests which would benefit patients (positive outcome), but reduce income for testing companies (negative outcome). The improvement in efficiency would likely not increase the income of many physicians as well. Also, given the ease at which information can be exchanged between health IT systems, patients whose physicians use them may feel that their privacy is more at risk than if paper records were used (Wikipedia n.d.)."

Figure 24.2 is a cartoonists view of what paper records look like hinting at the usefulness of storing medical records electronically.

In the document the information base for progress (Blumenthal et al. 2006) the section "Financial Barriers" (pp. 42 and 43) notes that:

> Financial barriers have a significant influence on HIT adoption. These barriers can be best understood as "twins:" the high cost of HIT systems; and provider uncertainty regarding the value they will derive from adoption in the form of return on investment. Stated another way, many providers do not perceive that there is a business case for HIT acquisition and use. They argue that the absence of a business case stems from a form of market failure within the HIT sector: current dysfunctional market dynamics and incentive structures do not work efficiently and effectively to realize the societal benefits of HIT. There are several reasons for this market failure. The first is that economic incentives in the health care industry generally do not reward good performance, reducing the motivation of self-interested health care actors to acquire HIT and compete more effectively. Often, health care compensation arrangements reward poor performance. Inefficient and sub-optimal care, for example, can generate more visits, tests and procedures and thus more revenue for providers. At a minimum, this reduces incentives for physicians and others to invest in systems to improve performance. Making matters worse, the purchasers of HIT—mostly doctors and hospitals—would capture only a small fraction of HIT's potential economic benefits. It has been estimated that as much as 80 percent of the potential savings generated through HIT inure to insurers and health care group purchasers, including the federal government, in the form of lower premiums and enhanced worker productivity.

"We just got an update to the user manual for our Electronic Medical Record system. Where do you want it?"

**Fig. 24.2** Illustrating the complexity of health records. (Reproduced under license from the creator)

The start-up costs for EMR implementation are significant. Figures of $50,000–$70,000/year for a physician practice were quoted in 2002 (Menachemi and Collum 2011) noting that costs do decrease with increased adoption. Vested parties such as EMR software systems vendors might tend to downplay the startup costs as well. However, it is found that systems' costs (when hospitals are included) tend to lower overall record managements costs and that significant ongoing saving can be expected as well as reduction in erroneous diagnoses and procedural errors can be realized.

Software today is increasingly complex and with the complexity there is an increase in unexpected errors. Furthermore software can be difficult to use, especially if the interface is poorly designed. Poor interface design can also lead to significant error when using the software. Although there are some standards for the encoding and storage of medical data the lack of a ubiquitous common terminology for EMR is a barrier to adoption.

In (Roehrs et al. 2017) the relationship between Personal health record (PHR) and electronic health record (EHR) relationships is depicted in Fig. 24.3 as:

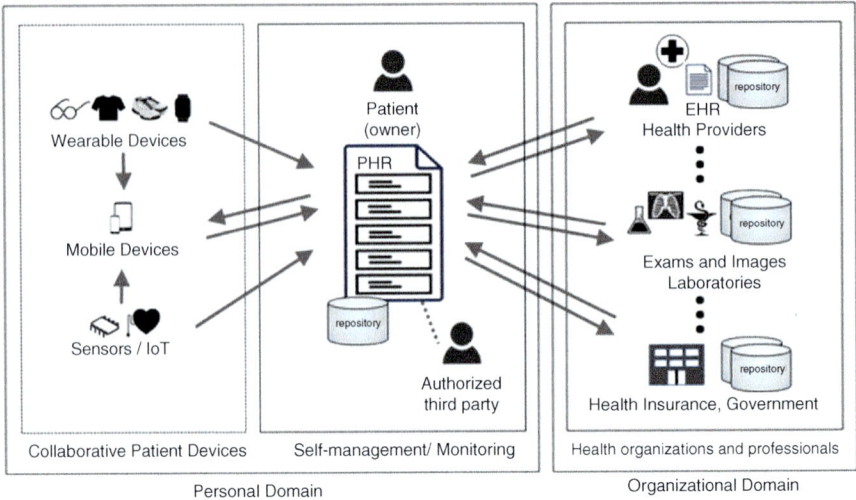

**Fig. 24.3** Connecting devices, records and providers to patients. (Reproduced under the Creative Commons Attribution License (http://creativecommons.org/licenses/by/2.0/))

**Table 24.1** PHR provider categories

| PHR architecture | Steady state net value (S/year, billion) | # of users per single PHR installation to break even |
|---|---|---|
| Provider-Tethered | −29 | 59,000 |
| Payer-Tethered | 11 | 62,000 |
| Third-Party | 11 | 47,000,000 |
| Interoperable | 19 | 52,000 |

Table available in the public domain

## 24.1   PHR Scenario

In this scenario the provider of PHR systems is delineated in Table 24.1 for the following four categories and for each category the steady state cost is estimated (Kaelber and Pan 2008):

A more detailed breakdown for each of the categories and PHR components is also provided below in Table 24.2 (Shah et al. 2008):

The above difficulties experienced for the implementation of are magnified for a PHR. A few examples illustrate this: Suppose US decides to implement a PHR system that is consistent across the 52 states (rather unlikely in the short term). A person travelling to Germany carrying his/her PHR might find that her/his PHR data cannot be read or if read, not understood due to differences in data formats, terminology, etc.

**Table 24.2** PHR costs by category

PHR single installation total costs acquisition and annual costs by architecture

| PHR component | Provider-Tethered ($) | | Payer-Tethered ($) | | Third-Party ($) | | Interoperable ($) | |
|---|---|---|---|---|---|---|---|---|
| | Acquisition | Annual | Acquisition | Annual | Acquisition | Annual | Acquisition | Annual |
| Clinical data repositories | $0 | $0 | $0 | $0 | $400,000 | $100,000 | $400,000 | $100,000 |
| Client user authentication | $95,000 | $14,000 | $95,000 | $14,000 | $95,000 | $14,000 | $95,000 | $14,000 |
| Core data user interface | $450,000 | $90,000 | $450,000 | $90,000 | $450,000 | $90,000 | $450,000 | $90,000 |
| Data center | $1,700,000 | $930,000 | $1,700,000 | $930,000 | $1,700,000 | $930,000 | $1,700,000 | $930,000 |
| Doctor matching | $0 | $0 | $0 | $0 | $0 | $57,000 | $0 | $57,000 |
| Interfaces | $40,000 | $8000 | $20,000 | $4000 | $6,600,000,000 | $1,300,000,000 | $250,000 | $50,000 |
| Medication matching | $0 | $0 | $0 | $0 | $0 | $17,000 | $0 | $17,000 |
| Network connectivity | $0 | $1000 | $0 | $1000 | $0 | $1000 | $0 | $1000 |
| Patient matching | $0 | $0 | $0 | $0 | $67,000 | $125,000 | $67,000 | $130,000 |
| PHR data repository | $0 | $25,000 | $0 | $25,000 | $0 | $25,000 | $0 | $25,000 |
| Results answer matching | $0 | $0 | $0 | $0 | $17,000 | $15,000 | $17,000 | $15,000 |
| Results name matching | $0 | $0 | $0 | $0 | $0 | $460,000 | $0 | $460,000 |
| User support | $0 | $2,700,000 | $0 | $2,700,000 | $0 | $2,700,000 | $0 | $2,700,000 |
| Secure messaging | $50,000 | $10,000 | $50,000 | $10,000 | $50,000 | $10,000 | $50,000 | $10,000 |
| Total cost | $2,300,000 | $3,800,000 | $2,300,000 | $3,800,000 | $6,600,000,000 | $1,300,000,000 | $3,000,000 | $4,600,000 |
| Single application cost | $450,000 | $90,000 | $450,000 | $90,000 | $450,000 | $90,000 | $450,000 | $90,000 |
| Total cost w/application[a] | $2,800,000 | $3,900,000 | $2,800,000 | $3,900,000 | $6,600,000,000 | $1,300,000,000 | $3,400,000 | $4,700,000 |
| Cost per user[b] | $3 | $4 | $3 | $4 | $6,600 | $1,300 | $3 | $5 |

[a]Numbers may be off due to rounding
[b]Assuming one million users
Table available in the public domain

On a positive note, there are a number of PHR benefits. Following is a list of PHR benefits shared by the Office of the National Coordinator for Health Information Technology in the US:

- Improve Patient Engagement. Much of what patients do for their health happens outside the clinical settings. When patients can track their health over time and have information and tools to manage their health, they can be more engaged in their health and health care.
- Coordinate and Combine Information from Multiple Providers. PHRs can promote better health care by helping patients manage information from various providers and improve care coordination.
- Help to Ensure Patient Information is Available. Online PHRs can ensure patients' information is available in emergencies and when your patients are traveling.
- Reduce Administrative Costs. Your organization can reduce administrative costs by using a PHR to provide patients with easy access to electronic prescription refill and appointment scheduling applications. With PHRs, your staff can spend less time searching for patient-requested information and responding to patient questions.
- Enhance Provider—Patient Communication. Many PHRs allow direct and secure communication between patients and providers. PHRs can make communicating with patients faster and easier. With open lines of communication, parties can be informed and intervene earlier if health problems arise, and improve the provider—patient relationship.
- Encourage Family Health Management. Having a system for tracking and updating health care information can help caregivers—such as those caring for young children, elderly parents, or spouses—manage patient's care and coordinate their efforts to improve health care quality (Menachemi and Collum 2011).

It is believed that the above mentioned benefits tend to reduce overall costs, especially in the long run (Tang et al. 2006).

## 24.2 Conclusion

Implementation of EHRs and PHRs can be costly. Costs pertain to various aspects, such as software design, standardization, user training, maintenance, privacy, and legal issues. Despite the costs incurred by the implementation of EHRs and PHRs, it is believed that benefits achived by their implementation are worth the cost.

# References

Blumenthal D, DesRoches CM, Donelan K, Ferris T, Jha A, Kaushal R, Rao S, Rosenbaum S. Health Information technology in the United States: The information base for progress. The Robert wood Johnson Foundation. 2006. https://hsrc.himmelfarb.gwu.edu/cgi/viewcontent. cgi?article=1473&context=sphhs_policy_facpubs. Downloaded 27 May 2019.

Canadian Medical Protection Association. Electronic records handbook. https://www.cmpa-acpm. ca/en/advice-publications/browse-articles/2014/electronic-records-handbook. Downloaded 29 May 2018.

HealthIT.gov. What are the advantages of electronic health records? Official Website of the Office of the National Coordinator for Health Information Technology (ONC). 2019. https://www. healthit.gov/faq/what-are-the-advantages-electronic-health-records. Downloaded 27 May 2019.

Kaelber D, Pan EC. The value of personal health record (PHR) systems. AMIA Annu Symp Proc. 2008;2008:343–7.

Luo G, Tang C, Thomas SB. Intelligent personal health record: experience and open issues. J Med Syst. 2012;36(4):2111–28.

Markle Foundation. Connecting for health. A public-private collaborative. The personal health working group final report, July 1, 2003. https://web.archive.org/web/20070104212409/http:// www.connectingforhealth.org/resources/final_phwg_report1.pdf.

Menachemi N, Collum TH. Benefits and drawbacks of electronic health record systems. Risk Manag Healthc Policy. 2011;4:47–55.

Roehrs A, da Costa CA, da Rosa Righi R, de Oliveira KSF. Personal health records: a systematic literature review. Journal of Medical Internet Research. 2017;19(1):e13.

Shah S, Kaelber DC, Vincent A, Pan EC, Johnston D, Middleton B. A cost model for personal health records (PHRs). AMIA Annu Symp Proc. 2008;2008:657–61.

Silva BMC, Rodrigues JJ, de la Torre Díez I, López-Coronado M, Saleem K. Mobile-health: a review of current state in 2015. Journal of Biomedical Informatics. 2015;56:265–72.

Tang PC, Ash JS, Bates DW, Overhage JM, Sands DZ. Personal health records: definitions, benefits, and strategies for overcoming barriers to adoption. J Am Med Inform Assoc. 2006;13(2):121–6.

Wikipedia. Electronic health record. https://en.wikipedia.org/wiki/Electronic_health_record.

# Chapter 25
# Threats and Barriers

Egondu R. Onyejekwe and Cory L. Hall

**Abstract** The establishment of a ubiquitous and authoritative portable health record for patients must overcome significant threats and barriers. The primary threat comes from the Electronic Health Record (EHR) industry that has a vested interest in maintaining the status quo. The primary barrier is finding an economic model that will support the creation and maintenance of an independent and transportable health record and the supporting infrastructure. On the other hand, having a portable health record will lower the cost of entry to the market, increase competition, and accelerate innovation. But, it will take either government or philanthropic intervention to change the course and establish an independent authoritative health record.

**Keywords** Portable health record · Economic model · Amazon · Berkshire Hathaway · JPMorgan chase · Software vendors · Continuity of care document (CCD) · Continuity of care record (CCR) · Clinical document architecture (CDA) · Consolidated-clinical document architecture (C-CDA) · Government intervention · Philanthropic intervention · Atul Gawande

## 25.1 Introduction

A truly portable health record is an authoritative health record that exists independent of the systems and processes that contribute to that record. All health care providers must accept that the information in that record is more reliable and comprehensive than any other single source of information. For example, a lab result in

E. R. Onyejekwe
Public Health, Health Administration, College of Health Sciences, Walden University,
Minneapolis, MN, USA
e-mail: Egondu.onyejekwe@mail.waldenu.edu

C. L. Hall (✉)
Department of Information Technology, University of Miami Health System,
Miramar, FL, USA

this record must be trusted by all providers based on the cryptographically certified credentials of the contributing laboratory.

Today's EHRs are not portable, they are predominately maintained by an ever-decreasing number of software vendors who have a strangle hold on the market. This market is dominated by two primary competitors Cerner and Epic, each with very different approaches to creating and maintaining their own proprietary version of a health record. Because of the proprietary nature of these health records, information transferred between vendors is dependent on a standard intermediate form of the health record called a Continuity of Care Document (CCD).

The Continuity of Care Document (CCD) is an electronic document exchange standard derived as a compromise by two standards groups, American Society for Testing and Materials (ASTM) International and Health Level 7 (HL7). These standards are used by Healthcare Delivery Organizations (HDO) to share summaries of patient care information for patients transferring between HDOs or receiving care from more than one HDO. The patient summary contains the most commonly needed information about healthcare provided by an organization to a specific patient. Such information must be in a form that can be interpreted by all participating computer applications, including web browsers, electronic medical record (EMR) and electronic health record (EHR) software. The specific content and scope of the CCD is built within a broader HL7 standard called Clinical Document Architecture (CDA). The CCD extends the concept established by ASTM called the Continuity of Care Record (CCR). Both the CCD and CCR use a general data structure standard, used extensively on the internet, called eXtensible Markup Language (XML). HL7 maintains that the CCD standard is an implementation of the CCR standard; others suggest that the CCD standard competes with the Continuity of Care Record standard. The CCD is the standard commonly used to exchange information between organizations because it is part of a much more comprehensive set of standards maintained by HL7. HL7 is also used to communicate between disparate electronic systems within an HDO, such as between a Laboratory Information System (LIS) and an EHR/EMR.

Transferring health record information between organizations using the CCD is like the game of telephone, where information propagated between organizations through the CCD sacrifices precision for portability. Current software vendors have a vested interest in keeping the record proprietary and maintaining control of the market. If a more portable record became the authoritative record, it would allow a wide variety of competing software vendors to erode the value of the proprietary record formats supported by the leading vendors. These vendors will do anything possible to maintain their current market dominance. Having a portable health record will lower the cost of entry to the market, increase competition and accelerate innovation.

The current EHR market creates an economic barrier and it will take either government or philanthropic intervention to make a change in a favorable direction. The recent announcement by Jeff Bezos (Amazon), Warrant Buffet (Berkshire Hathaway) and Jamie Diamond (JPMorgan Chase) to collaborate to "Fix" healthcare could signal an opportunity to be seized, to make portable health records a part of that effort. The required investment would be used to create a baseline standard portable health record, methods for collaborative non-repudiable contribution, and maintenance of the supporting infrastructure.

## 25.2  Portability of the Current EHR

A truly portable health record is an authoritative health record that exists independent of the systems and processes that contribute to that record. All health care providers must accept that information in that record is more reliable and comprehensive than any other single source of information. For example, a lab result in this record must be trusted by all providers based on the cryptographically certified credentials of the contributing laboratory.

Today's EHRs are not portable. They are predominately maintained by an ever-decreasing number of software vendors who have a strangle hold on the market. This market is dominated by two primary competitors Cerner and Epic. Each with very different approaches to creating and maintaining their own proprietary version of a health record. Because of the proprietary nature of the health record, data transferred between these vendors are dependent on a standard intermediate form of the health record called a Continuity of Care Document (CCD).

As indicated in the introduction, the CCD is an XML-based standard that marries the best of HL7 technologies to the richness of the clinical data representation provided by ASTM's Continuity of Care Record (CCR) format, without interrupting existing data flows. Meaningful Use Stage 1 thus allowed two different types of documents to be sent. They include: Continuity of Care Record (CCR) or the Continuity of Care Document (CCD). CCD was mandated for Meaningful Use Stage one. These two standards, CCR and CCD, plus the CDA resulted in a total of three standards.

- Continuity of Care Record (CCR) is an ASTM standard
- Continuity of Care Document (CCD)  is an HL7 standard and part of the CDA (Clinical Document Architecture) family of standards
- Clinical Document Architecture (CDA)

These constitute three Cs of healthcare. Consolidated CDA (C-CDA) emerged with the final rule for Meaningful Use Stage 2 in 2012 to offset the different difficulties associated with CCD.

While the history of the origins of the three Cs are important, they are less relevant to this discussion. What is pertinent is that the health IT industry ended up with two valid formats instead of one for Meaningful Use Stage 1. But then in Meaningful Use stage 2, those two formats were reduced to one.

## 25.3    A Brief Overview of XML

It is important to briefly address the Extensible Markup Language (XML). XML is a standard used to describe and structure data. It provides a flexible way to define information and electronically share information though both the public Internet and corporate networks. The World Wide Web Consortium (W3C) formally recommends XML. HTML used across the internet is a direct derivative of XML. They are both based on Standard Generalized Markup Language (SGML) and contain "markup" character sequences that describe the contents of an electronic document in an unambiguous way. An early version of HTML, implemented on web browsers, supported nonstandard XML shortcuts that helped reduce the size and complexity of early web pages. These shortcuts introduced ambiguities that eventually caused problems on the internet. Although older HTML is still supported for backward compatibility, newer HTML standards XHTML and HTML5 more strictly adhere to the XML standard.

A Simple XML document:

```
<home>
    <address>
        <street>763 Park Lane</street>
        <city>Mayberry</city>
        <state>Kansas</state>
        <postcode>34567</postcode>
    </address>
    <note>Restricted access, guarded community</note>
</home>
```

## 25.4    The Continuity of Care Document and the Consolidated Clinical Document Architecture

Transferring health record information between systems using the CCD is also like the game of telephone, where information propagated between systems through the CCD sacrifices precision for portability. The CCD does not solve many of the data portability problems. At its foundation, the CCD was always intended to be a summary of care provided elsewhere. Epic calls its CCD based data sharing process "Care Elsewhere". CCD is not comprehensive enough to serve as a primary record of care. To address some of these concerns, the Healthcare Information Technology Standards Panel (HITSP) decided to provide additional guidance in the use of the CCD standard, suggesting that certain sections adhere to a more specific format within the CCD. However, it did not require that the guidance be adhered to.

Meaningful Use Stage 2 (MU Stage 2) began in fiscal year 2014 for eligible hospitals/critical access hospitals and eligible professionals and specified Consolidated-Clinical Document Architecture (C-CDA).

While CDA, the base standard, provides a common architecture, coding, semantic framework, and markup language for the creation of electronic clinical documents, the C-CDA in general, squeezes nine different file types, progress notes, clinical summaries, consult notes, and the CCD into one file, with consistent headers.

The two major objectives of the C-CDA were: (a) to eliminate multiple sources of truth; and (b) to establish a mandate to be used freely. The first objective was achieved by eliminating multiple specification documents. The second was achieved by having one implementation guide at no cost.

CMS' August 2015 paper on Stage 2 Eligible Professional Meaningful Use Core Measures provided information on the transition of care by eligible professionals (EP). Besides their definition of transition of care: "The movement of a patient from one setting of care (hospital, ambulatory primary care practice, ambulatory, specialty care practice, long-term care, home health, rehabilitation facility) to another", they provided the minimum requirements for such transactions. Among them are that the summary of care record must at a minimum include all transitions of care and referrals the EP has ordered. Such a summary of the care record—in essence, the electronic health record (EHR), must include the following elements:

- Patient name
- Referring or transitioning provider's name and office contact information (EP only)
- Procedures
- Encounter diagnosis
- Immunizations
- Laboratory test results
- Vital signs (height, weight, blood pressure, BMI)
- Smoking status
- Functional status, including activities of daily living, cognitive and disability status
- Demographic information (preferred language, sex, race, ethnicity, date of birth)
- Care plan field, including goals and instructions
- Care team including the primary care provider of record and any additional known care team members beyond the referring or transitioning provider and the receiving provider
- Reason for referral
- Current problem list (EPs may also include historical problems at their discretion).
- Current medication list
- Current medication allergy list

This means that all Electronic Health Records or EHRs that were meaningful use stage 2 certified support C-CDA. C-CDA is an architecture used to create templates and documents for clinical records. The primary function of the C-CDA is to standardize the content and structure for clinical care summaries. C-CDA documents are coded in XML and XHTML and are made of these parts:

- Header—(enables exchange of clinical documents within and across institutions)
- Body—(contains the clinical report and can contain unstructured or structured content, in one or more Sections)

    (a) Section(s)—(may contain Allergies, Meds, Problems, Immunizations, Vital Signs, etc.)
    (b) Narrative block—("human-readable" part of a CDA document)
    (c) Entries (0 to maximum)—(structured "machine-readable" content for further computer processing)

Starting December 2011, the C-CDA was focused on making CDA Templates (CCD being one of its templates). It was adopted by the United States through Meaningful Use efforts. Some experts maintain that C-CDA will make it incrementally easier to achieve "international interoperability, and deliver on the promise of persistence, stewardship, potential authentication, context, wholeness and human readability, as originally promised by CDA in 1999."

It is a single standard for communicating summary of care records, enabling the sharing of clinical care information in the most common transition of care scenarios. Among such scenarios are: inpatient-to-outpatient, primary care physician (PCP)-to-specialist, provider-to-patient and provider-to-ACO. C-CDA is also believed to facilitate easier EHR-to-EHR sharing of patient care records.

The C-CDA accommodates more than summaries and snapshot stories for patients. It is based on the HL7 Reference Information Model (RIM), it provides flexibility to accommodate user-defined fields, and can store complete documents, structured data, and multimedia. It was approved by ANSI in 2010. At its core is the CDA which introduces the concept of incremental semantic interoperability, that can link elements of care. The minimal CDA is a small number of XML-formatted fields (such as provider name, document type, document identifier, and so on) and a body which can be any commonly-used document type including "pdf" (Adobe Acrobat), "doc" (Microsoft Word) or a "dicom" image. C-CDA has two separate components:

1. Human readable components found in section/text, and
2. Machine readable components found in section/entry

C-CDA documents can be encoded with the full power of the HL7 Reference Information Model (RIM) and controlled vocabularies such as LOINC, SNOMED, ICD, CPT, RxNorm etc. C-CDA is clearly the next step in the evolution of interoperability because it allows substantially easier point to point connection, i.e. you don't have to go through a post office model like state/regional Health Information Exchanges (HIE), and it allows for much better patient engagement with human readability of complex data and patterns of data.

Is this an answer to porting EHRs? That remains to be seen.

Other experts argue that while on paper, this reads like the ultimate solution, the standard still requires vendors to understand the intricacies of each file type. As with all standards therefore, there are still pros and cons for each format.

Burgher explored the strengths and weaknesses of different data formats in a blog. These are provided below.

The pros for CCR include the following:

- Relatively robust data set
- Theoretically a one-stop shop for sending entire PHM record
- Good for historical transfer but incremental updates require sending everything again

The cons for CCR include:

- Almost no rules for data presentation (e.g., medication name only versus National Drug Code)
- One-stop shop approach generally includes large files that can bog down transfer. Pathways are generally optimized for either a large number of small files (like HL7) or for large files
- Potential for errors increases due to file size

The pros for CCD include:

- Standard that married the best of HL7 and CCR
- Includes content and structure
- First step in formalizing the content that should be sent

The cons for CCD include:

- Doesn't require accepted coding systems (NDC, LOINC)
- Data model for CCD is complex
- Although human readable, involves is a steep learning curve because the structure of CCD is based on CDA, which, in turn, is based on Reference Information Model (RIM); a large, pictorial representation of the HL7 clinical data)

The pros for C-CDA include:

- Attempts to consolidate the various CCD flavors into one
- Adds more recommendations for what coding systems to use (not enforced)
- Changed what sections are required to further constrain data

The cons for C-CDA include:

- Recommended coding systems not enforced
- Relatively complex to read, but not as difficult as X12
- Still based on RIM
- Pursuit of objectives within 2 of the 7 MU2 categories involve using Certified EHR Technology that has C-CDA capabilities (Care Coordination & Patient Engagement)

The goal of implementing C-CDA in Meaningful Use Stage 2 was mainly to simplify the learning curve required in understanding the structure of the required documents. The Consolidation of all documentation under one implementation guide plus the use of templates is worthy of acknowledgement, at least as a path towards achieving this goal. However, with the introduction of complexity that was added with four different types of summary documents for the different requirements of Meaningful Use the systems were enamored. The tradeoff was that this was deemed necessary to provide only the required clinical information necessary for each use case.

C-CDA has its critics. In October 2014, the American Medical Association released a document expressing concerns over C-CDA and its ability to advance interoperability—that ability of two or more systems or components to share information, which is at the core of portability of electronic health records (EHR). The document opined that the Office of the National Coordinator for Health Information Technology (ONC)—mandated the use of C-CDA in Stage 2 even though it had "very little real-world testing, nor was it balloted or approved for standardization by the Standards Developing Organization (HL7)" that is accredited by the American National Standards Institute. Because of this non HL7 approval it had "wild variation in technology versioning."

These competing standards were created to shuttle information from one electronic system to another, replicating information rather than establishing a shared authoritative medical record. Each replication sacrificing fidelity for practicality. The result is that no one system will have a comprehensive and accurate history of the patient's past care and needs related to their ongoing medical problems. It seems clear that the final solution is not how information is shared but how do we create a common shared record of care managed by the patient, that no one organization owns and no one vendor controls.

## 25.5   The Economic Barrier of the EHR Market

The economic barrier created by the current EHR market is enormous. It will take either government or philanthropic intervention to change the course and establish an independent authoritative health record. Any commercial enterprise is destined to fail by pursuing direct economic gain. An analogy can be drawn from our national power grid and highway system. Both these systems would never work unless a basic set of standards were established and a public investment made. For the highways it was a national interstate highway system that established the back bone. For the power grid it was a common set of voltages, alternating current frequency, efficient transformers and baseline power generation.

The Press Release on January 30th, 2018, that three corporate behemoths Amazon, Berkshire Hathaway, and JPMorgan Chase would partner to provide their combined 1.1 million US employees healthcare sent shock waves throughout the Industry. Their major concerns are that the U.S. healthcare costs were rising too fast and holding back economic growth. They will essentially launch an independent healthcare operation that's intended to be free from profit-making incentives. They said they would use big-data analysis and other high-tech tools to improve care and cut waste.

All major news networks (CNBC, Reuters', CNN, NPR, USA Today WSJ, and NY Times among others) and major Corporations (including Bloomberg) addressed the Press Release. While these three companies are currently focusing on their joint staff at the present time, it is predicted to set disruptive waves throughout the health industry. Bloomberg maintains that the "Amazon-Berkshire-JPMorgan collaboration will likely pressure profits for middlemen in the health-care supply chain." CNBC surmised that:

- Essentially, Amazon, Berkshire Hathaway and JPMorgan announce a partnership to cut health costs and improve services for employees.
- The idea is to create a company that would be "free from profit-making incentives."
- News of the deal slammed suppliers in the industry including Express Scripts, Cigna, CVS, United Health and Aetna.

Noteworthy is that the new company's goal will at first "target technology solutions to simplify the health-care system. So, the new company will focus on technological solutions to provide coverage for their U.S. employees at a lower cost. As much as the details remained sketchy, their sheer size will help bring the scale and resources to tackle the issue of technology solutions more effectively and more efficiently.

On June 20th, 2018, Atul Gawande was Named CEO Of Health Venture by Amazon, Berkshire Hathaway And JPMorgan. Atul Gawande is a prominent surgeon, author and checklist-evangelist, and some sources opine that these attributes signal ambitious plans for tackling both costs and quality of healthcare.

Gawande started as CEO July 9, 2018, and the venture, still an un-named company is headquartered in Boston. Most of the healthcare community signal both skeptical and hopeful vibes for the venture and Gawande's appointment. He has the unenviable goal of creating "innovations in a burdened and dangerously large healthcare industry, lowering the cost of care while improving health outcomes." Analysts see the choice of Gawande as a reflection of the company's plans to focus on the entire healthcare system. So, they will not be limited to just curbing prescription drug costs. Buffett described U.S. healthcare costs as a "tapeworm" on American businesses, that hurts their ability to compete with rivals in other countries. So, the goal, to him is to challenge the entire healthcare industry, not individual segments.

Gawande's selection to lead this effort while puzzling to a few health policy experts since he is not known to be an expert on the business and financing of health care, is heralded by some because he has focused more on improving the delivery of care by hospitals and doctors, as well as other policy issues. Gawande has been known as a spokesperson for change and new ideas in the healthcare system. Combined with his training as a doctor and healthcare academic, there is a chance that he would be grounded and realistic about making the required change. The big and open question is how his ideas will translate into change and how they may address the need for a portable health record.

## 25.6 Conclusion

The creation of common health record will help the industry focus on meeting the needs of the individual patient rather than on the process of managing conflicting agendas and incentives. Any organization involved with the patient's care can interrogate and contribute to the common record based on need and authorization. This common record would be a foundation for research, quality improvement, coordination of care, multidisciplinary disease management, and cost analysis. This would not be beyond the reaches of Dr. Gawande and his team. They should set up an independent organization charged with creating and managing the health records for all patients they cover. Then, they should force providers to use and contribute to that record when providing care for patients covered by their system. The required investment would be used to create a common portable health record, methods for non-repudiable contribution, patient control and maintenance of the supporting infrastructure.

# Part VIII
# Conclusion

# Chapter 26
# Conclusion: Health Record Portability

Egondu R. Onyejekwe and Cory L. Hall

**Abstract** We conclude this book by revisiting Health Record Portability. The future of the provision of healthcare in the United States (US) of America is not dependent on any single factor. Several—internal and external factors—such as skyrocketing costs, personalized medicine, regenerative medicine, evidence-based medicine, gene therapy, robotic surgery, artificial organs, technology and their applications—all affect the delivery of healthcare in the US. The social impact will be enormous as society struggles to align priorities and deal with socioeconomic inequalities. Health Record Portability involves spacetime because the movement of health records or parts thereof expend time and space for the interactions. Many chapters of this book try to present the different aspects of the healthcare ecosystem, which includes the electronic health record (EHR) or electronic medical record (EMR). Integral to these is how EHR or EMR is moved from point A to point B. As articulated in the introductory paragraph, frequency and magnitude of spacetime and the interactions of space and time at every encounter of such a motion increase cost. Some emerging technologies are envisaged to provide one or more directions that conserve/s time and space and thus reduces the spiraling cost/s of a fragmented US healthcare system. As indicating earlier, every porting (movement) of an electronic health (medical record) involves time and space, the interaction of which results in a specified cost. But the diversity and variations of today's EHRs (EMRs) make it difficult to pigeonhole them inside any standardized form. The costs of processing them have become incalculable, and the healthcare costs in the U.S. are spiraling out of control. Moreover, notions of EMR/EHR vary across the globe, but the US definitions are a far cry from the European definitions. Yet, the US EMR/EHR are governed by certain laws, whose limits are unfortunately worrisome. In many cases patients find themselves carried along the currents and eddies of their

E. R. Onyejekwe (✉)
Public Health, Health Administration, College of Health Sciences, Walden University, Minneapolis, MN, USA
e-mail: Egondu.onyejekwe@mail.waldenu.edu

C. L. Hall
Department of Information Technology, University of Miami Health System, Miramar, FL, USA

© Springer Nature Switzerland AG 2019
E. R. Onyejekwe et al. (eds.), *Portable Health Records in a Mobile Society,*
Health Informatics, https://doi.org/10.1007/978-3-030-19937-1_26

to medical devices; creating an opportunity for the advancement of the use immunotherapy for cancer treatment; turning the electronic health record into a reliable risk predictor; allowing the monitoring of health through wearables and personal devices; converting smartphone selfies into powerful diagnostic tools; and revolutionizing clinical decision making with AI at the bedside. Additionally, AI could be the doctor of the future, where its use as the stethoscope is one that would be used in patient diagnosis. There are many other AI trends that have been addressed in that chapter.

2. *Block Chain (BC)* (Health Data Management 2016c)—this is a technology that is being adapted for the improvement of electronic health records. BC is a log of transactions that is replicated and distributed across multiple decentralized locations. So, it offers a secure, high integrity, "neutral" 3rd party mechanism *"for knowing what data is where and precisely how it is changing over time."* A whole chapter is devoted to this technology because of its current relevance to EHR.

3. *The Cloud* (Health Data Management 2016d)—it is estimated that about 83% of healthcare industry uses the Cloud. Because it provides a flexible technology architecture, it is easy to use and can bridge the gaps among the technologies that are already in use. Its major problems lie with security of health data.

4. Consumer—facing technology (Health Data Management 2016e)—these are more rampant with mobile apps—and include an expansion of digital communication and available patient data which will move the healthcare industry way beyond the conventional doctor's visit. West Monroe Partners found that 91% of the customers in healthcare do take advantage of the mobile apps when these are offered. They further stated that 80% of this group do prefer mobile to conventional office visit. Among the other tech initiative are: healthcare insurers, where about 70% of them currently offer rewards programs that harness data from consumer's health tracking devices and apps; and data analytics in healthcare which will advance to help patients in saving money. It is important here to understand and differentiate what useful data is versus data that is simply noise. Useful data will invariably encourage and drive patient action.

5. Disease Management Technology (DMT) (Health Data Management 2016f)— Disease management technology is crucial for healthcare organizations because it facilitates the speed and shortens the time to market. Healthcare organizations care are constantly under pressure to differentiate disease treatments, as well as include speed and agility. In totality, DMTs will provide "easily accessible complete views of stakeholder, which will in turn, enable proper orchestration of customer engagement which would then lead to the achievement of the commercial and R&D goals. There is an expansion of digital communication and available patient data which will move the healthcare industry way beyond the conventional doctor's visit.

6. EHR Improvement (Health Data Management 2016g)—As the ubiquity of Electronic Health Records (EHRs) present a catch-22 scenario, according to Brent Lang, CEO of Vocera Communications, EHRs have promoted the digitization of healthcare, but, they have also increased the burden on clinicians.

Incessant frustrations among clinicians that result from EHRs range from the lack of interoperability to the excessive documentation burdens placed on them. Brent Lang opined that: "EHR pain points and fatigue are new dynamics; any healthcare technology that addresses those issues will endure."

7. Imaging (Health Data Management 2016h)—Medical imaging involves several different technologies that provide different information about a particular area of the human body that is under study. It is used to diagnose, monitor, or treat medical conditions about body area being studied or treated. These parts could be related to possible disease, injury, or the effectiveness of medical treatment.

8. Internet of Things (Health Data Management 2016i)—The Internet of things (IoT) is expected to increase with the growing ability to connect medical devices and other gadgets to the Internet. With this growth come the security challenges. As indicated earlier both medical care and healthcare represent several attractive areas for the IoT applications including remote health monitoring, fitness programs, chronic diseases, elderly care, treatment and medication compliance both at home and by healthcare providers. Consequently, the different medical devices, sensors, and diagnostic and imaging devices can become the needed smart devices or objects that constitute a core part of the IoT.

    So, IoT-based healthcare services increase the quality of life and enrich the user's experience. As a consequence, for the healthcare providers, and for those using remote provisions, the IoT has the potential to reduce device downtime, correctly identify optimum times for replenishing supplies for various devices so they can run smoothly and continuously as well as provide for the efficient scheduling of limited resources, especially in resource and skill poor areas. Spacetime and the inherent increase in interactions are reduced, thereby reducing costs. Thus, the "IoT revolution is redesigning modern health care with promising technological, economic, and social prospects."

9. Interoperability (Health Data Management 2016j)—Interoperability is at the core of moving electronic health records (EHRs). The 21st Century Cures Act makes provisions for the EHRs. A provision in the Act seeks to improve interoperability, thereby improving patient care. Both semantics and syntactics are taken care off, so the record arrives with greater fidelity than currently occurs. Furthermore, more effort is expected in the information exchange initiatives as well as in the development and use of Fast Healthcare Interoperability Resources (FHIR) (Fast Healthcare Interoperability Resources 2017).

10. Telemedicine (Health Data Management 2016k)—Smith tried to distinguish telemedicine and telehealth in this brief titled "Telemedicine VS Telehealth: What's the Difference?" People in the industry use both terms interchangeably, but Smith draws the distinction below to separate them. This is a recap from a previous chapter.

    First, telehealth is a subset of E-Health. E-Health in turn, uses the internet and telecommunications to provide a vast array of services, such as the "delivery of health information, for health professionals and health consumers, education and training of health workers and health systems management (Chiron Health 2015)."

Telehealth, she opined, includes "a broad range of technologies and services to provide patient care and improve the healthcare delivery system as a whole (Chiron Health 2015)." Telehealth, thus, refers to both clinical services and nonclinical services such as remote non-clinical services, including provider training, administrative meetings, and continuing medical education, in addition to clinical services. The World Health Organization (WHO), indicates that telehealth also, includes, "surveillance, health promotion and public health functions (Chiron Health 2015)." Telehealth, is more encompassing and refers to a broader scope of remote healthcare services than does telemedicine, which refers specifically to remote clinical services.

Telemedicine, is simply a subset of telehealth that refers solely refers to the use of telecommunications technology to provide health care services and education over a distance (HealthIT.gov 2017). Additionally, while telemedicine involves the use of electronic communications and software to provide clinical services to patients without an in-person visit, telemedicine technology can often be used for follow-up visits, management of chronic conditions, medication management, specialist consultation among other clinical services that can be provided remotely through audio and video connections that are secured.

Finally, portable health records also imply that the Providers of services can share and exchange information on a patient's behalf. Because the health record will reside in the cloud or some secured place, only those authorized can access and download on their mobile devises, thereby offsetting security and privacy issues. The commonality of language (semantics and syntactics) will become crucial.

## 26.3   Health Technology Decisions

A deluge of technologies that are being applied to healthcare and those with potential applicability in the future of healthcare have been described. While they present several opportunities, they could also lead to decision paralysis for those who must make those choices.

How will clinical and business decision support be enabled using Portable Health Records? Clinical decision support is helping care providers make decisions about the care for individual patients based on an available clinical knowledge base (Norman 2018). Business decision support is helping administrators make informed decisions based on the analysis of the organizations specific mix of patients and a business knowledge base.

Cloud computing is moving at an accelerated speed (BMC Medical Informatics and Decision Making 2015). But, the applications so defined, don't quite meet the attributes required of cloud computing. They are moreover burdened by privacy and security issues.

The Internet of Things (IoT) seems so close to resolving the issues because it is virtually "a connected set of anyone, anything, anytime, anyplace, any service, and any network (Researchgate 2015)." It would therefore make time and place irrelevant for both provider and patient in the healthcare field. But then, what does it operate on—Cloud computing? Cloud computing when in association with the Internet of Things (IoT), nothing but obfuscation results (IGI Global 2018).

Telemedicine and imaging technologies have limited application areas and are considerably pricy. So are disease management technologies and electronic health records.

So, the best advice to a healthcare provider who aspires to invest in technology for the delivery of healthcare would be: "Buyer Be Ware". There is so much fuzziness in the field, many competing and complicating demands, and no strategic direction towards elimination of waste, reduction of costs, or more affordable healthcare—at least, not yet in the US!

Also, artificial intelligence seems like a viable option—but only to the extent that it is affordable (Bresnick 2018)!

Today's EMR/EHR is no closer to *portability,* than yesterday's was. It operates in the U.S. health care environment that is riddled with inefficiencies and high cost of delivery (HealthIT.gov 2018). It has no defined route/s for Access (in all its dimensions). It is hindered by various laws and legal underpinnings.

Above all though, *block chain* holds some promise, at least, it alleviates the problems of interoperability and holds the promise for security and privacy of data… except that it is still not a mature technology (HIMSS 17 2017)! It needs to address challenges that are outside healthcare, such as technical, organizational, and behavior challenges. Commoditization is also worth considering for the future.

## 26.4   Our Proposal

Throughout the topics addressed in this manuscript, our proposition remains the same: the reduction of spacetime interactions through—the creation of an authoritative digital record that exists independent of the delivery systems. This record would be collectively and directly maintained by care providers under the supervision of the patient or their designated agent. This record would be inherently portable.

This collection of essays tries to identify some of the most important issues and some of the most promising solutions to creating a truly portable health record. Our intent is not to create a comprehensive solution or recipe, but to stimulate thinking and hopefully establish a direction. One of the issues, how to create an authoritative record, exists independent of the organizations providing care and the electronic systems they use to manage their processes. We believe that technology can solve this problem using blockchain like processes that insure the integrity of an externally maintained health record (Zhang et al. 2016; Beck's Health IT and CIO Report 2016). Semantic consistency is another issue to be solved. Over many years of hard work a wide spectrum of nomenclature standards has been established, ready for adoption into a portable record. To a large extent the central issue is helping the healthcare delivery industry to transfer control over a patient's primary health record from isolated islands of corporate information to collective ownership, overseen by the patient or their authorized agent. This is similar to the creation of community power utilities and the national power grid, or local streets/highways and the national interstate system.

Availability, reliability and integrity of a portable health record are important to adoption and sustainability. Organizations providing emergency patient care need unhindered immediate access to the patient's medical record, irrespective of where care is provided. In most truly emergency situations patients receive care in an environment devoid of information from the historical patient medical record. In some cases, optimal care must be withheld because the care providers have incomplete knowledge, or a more expensive/scarce resource must be used. For example, the use of O-Negative blood for a patient with unknown blood type. Any portable health record must be trusted to be useful. Users of the portable health record must be comfortable that information in the record can be trusted, it must be complete and unadulterated. Adoption of a cryptographically secure and distributed storage model for portable health records must be developed to address these concerns.

Progress toward the goal of a distributed portable medical record will be impeded by market pressure unless supported by government intervention and/or philanthropic investment. This will be a long-term effort in an industry that evolves at an excruciating slow pace, driven by an out of balance risk reward system. The cost of failure in patient care, although justified in some cases, is so high that a very conservative and incremental change process is followed. On the other side the reward for relatively low risk change can be very high, as evidenced in recent medication pricing of epinephrine autoinjectors. Participants working toward an independent authoritative health record will need regulation, incentive and protection before progress can be made. It is our belief that the creation of a quasi-public government corporation like the Fannie Mae or philanthropic organization like the Kaiser Family Foundation or a combination thereof will be required to orchestrate development and implementation.

Health records are fundamentally private. In today's environment any information stored electronically, like a portable health record, is subject to being compromised. The information in this record must be secure but easily accessible to authorized individuals with specific needs. At the same time deidentified information should be available to researchers and public health officials. This means that a multi-stage encryption process needs to be employed. These stages will incrementally peel back layers and classes of information based on an authorization matrix provided by the patient or their agent. Public health officials and researchers should be able to interrogate the files for aggregated vaccination and disease occurrence information without direct access to individually identifiable information. Patient's or their agents should be in control of what type of information is made available, when that information is made available and have the ability to establish individually tailored access controls. This will require a cryptographically secure structure that will allow the information to be stored publicly and access requests approved privately through secure and private channels of communication.

Although other countries don't struggle with the same constraints on their healthcare systems as the United States, they too would benefit from an established and effective portable health record. The National Health System in the United Kingdom and the Provincial Health Systems in Canada have been struggling for years to create an effective national health record with varying success. It is possible that the World Health Organization would also find value in establishing an international health record standard.

What happened to the concepts of time and space presented in the Introductory chapter? They have moved! Time has moved, and space has also shifted only because the patient and the electronic health records have moved, too. Worse still they are constantly moving and due to accelerated interactions. As a consequence, costs continue to spiral. Is it still possible to capture the health records in an electronic mobile form that yields less costs? As has become obvious, the greater the interactions of space and time (spacetime) the more the costs escalate!

If, however, Artificial Intelligence tools such as machine learning are appropriately applied, the relevant health data can be captured even as the patient moves. Furthermore, if healthcare moves more towards commoditization, and if *BlockChain* technology application towards EHR/EMR are effected, the interaction of time and space would lessen and thus ease the cost (Deloitte 2018). Apparent areas for cost reductions would include: faster data exchanges due to interoperability; limited data exchange errors (both semantic and syntactic); security of data; and the privacy (U.S. Food and Drug Administration 2018) of the patient data. Consequently, and as stated in the introductory chapter, spacetime—the interplay of space and time— become both a necessary and a sufficient condition! This interplay would affect the cost of care, by streamlining it, reducing it, or completely eliminating it!

Additionally, and critical to all involved entities (patient and providers—including doctors, pharmacists, nurses and other health care providers) is the fact that the digital exchanges (porting) of the patient health information potentiates improvements in different areas, including: speed, quality, safety and cost of patient care. Spacetime is also conserved! In aggregate, the outcomes should provide rich data for data analytics—on individual, group, or societal level/s!

But until we arrive at the point where spacetime (the interactions are minimal, and costs take a downward spiral) we boldly propose the creation of an authoritative digital record that exists independent of the delivery systems. This record would be collectively and directly maintained by care providers under the supervision of the patient or their designated agent. This is the record that would be inherently portable!

It is our sincere hope that the work invested in this publication will be educational, stimulate debate, and promote efforts toward building a portable health record. Your engagement is desired!

# References

Beck's Health IT & CIO Report. 9 things to know about blockchain in healthcare. 2016. https://www.beckershospitalreview.com/healthcare-information-technology/9-things-to-know-about-blockchain-in-healthcare.html. Accessed 1 Nov 2018.

BMC Medical Informatics and Decision Making. A scoping review of cloud computing in health-care. 2015. https://bmcmedinformdecismak.biomedcentral.com/articles/10.1186/s12911-015-0145-7. Accessed 1 Nov 2018.

Bresnick J. Top 12 ways artificial intelligence will impact healthcare. 2018. https://healthitana-lytics.com/news/top-12-ways-artificial-intelligence-will-impact-healthcare. Accessed 1 Nov 2018.

Chiron Health. Telemedicine vs. telehealth: what's the difference? 2015. https://chironhealth.com/blog/telemedicine-vs-telehealth-whats-the-difference/. Accessed 1 Nov 2018.

Deloitte. Blockchain: opportunities for health care. 2018. https://www2.deloitte.com/us/en/pages/public-sector/articles/blockchain-opportunities-for-health-care.html. Accessed 1 Nov 2018.

Fast Healthcare Interoperability Resources. FHIR overview. 2017. https://www.hl7.org/fhir/over-view.html. Accessed 1 Nov 2018.

Health Data Management. 10 top healthcare information technology trends for 2017: new trends accelerate change within healthcare IT. 2016a. https://www.healthdatamanagement.com/list/top-healthcare-it-trends#slide-1. Accessed 1 Nov 2018.

Health Data Management. 10 top healthcare information technology trends for 2017: artificial intel-ligence. 2016b. https://www.healthdatamanagement.com/list/top-healthcare-it-trends#slide-2. Accessed 1 Nov 2018.

Health Data Management. 10 top healthcare information technology trends for 2017: block-chain. 2016c. https://www.healthdatamanagement.com/list/top-healthcare-it-trends#slide-3. Accessed 1 Nov 2018.

Health Data Management. 10 top healthcare information technology trends for 2017: the cloud. 2016d. https://www.healthdatamanagement.com/list/top-healthcare-it-trends#slide-4. Accessed 1 Nov 2018.

Health Data Management. 10 top healthcare information technology trends for 2017: consumer-facing technology. 2016e. https://www.healthdatamanagement.com/list/top-healthcare-it-trends#slide-5. Accessed 1 Nov 2018.

Health Data Management. 10 top healthcare information technology trends for 2017: disease management technology. 2016f. https://www.healthdatamanagement.com/list/top-healthcare-it-trends#slide-6. Accessed 1 Nov 2018.

Health Data Management. 10 top healthcare information technology trends for 2017: EHR improve-ment. 2016g. https://www.healthdatamanagement.com/list/top-healthcare-it-trends#slide-7. Accessed 1 Nov 2018.

Health Data Management. 10 top healthcare information technology trends for 2017: imag-ing. 2016h. https://www.healthdatamanagement.com/list/top-healthcare-it-trends#slide-8. Accessed 1 Nov 2018.

Health Data Management. 10 top healthcare information technology trends for 2017: internet of things. 2016i. https://www.healthdatamanagement.com/list/top-healthcare-it-trends#slide-9. Accessed 1 Nov 2018.

Health Data Management. 10 top healthcare information technology trends for 2017: interoper-ability. 2016j. https://www.healthdatamanagement.com/list/top-healthcare-it-trends#slide-10. Accessed 1 Nov 2018.

Health Data Management. 10 top healthcare information technology trends for 2017: telemedi-cine. 2016k. https://www.healthdatamanagement.com/list/top-healthcare-it-trends#slide-11. Accessed 1 Nov 2018.

HealthIT.gov. Telemedicine and telehealth. 2017. https://www.healthit.gov/topic/health-it-initia-tives/telemedicine-and-telehealth. Accessed 1 Nov 2018.

HealthIT.gov. Medical practice efficiencies & cost savings. 2018. https://www.healthit.gov/topic/health-it-basics/medical-practice-efficiencies-cost-savings. Accessed 1 Nov 2018.

HIMSS 17. The interoperability imperative: value-based care depends on health information exchange. 2017. http://www.healthcareitnews.com/himss-infocus/interoperability?utm_campaign=himss-infocus&utm_medium=text_ad&utm_source=himssorg&utm_term=infocus-interop. Accessed 1 Nov 2018.

IGI Global. Centralized fog computing security platform for IoT and cloud in healthcare system. 2018. https://www.igi-global.com/chapter/centralized-fog-computing-security-platform-for-iot-and-cloud-in-healthcare-system/187898. Accessed 1 Nov 2018.

Norman A. Your future doctor may not be human. This is the rise of AI in medicine. 2018. https://futurism.com/ai-medicine-doctor/. Accessed 1 Nov 2018.

Researchgate. The internet of things for health care: a comprehensive survey. 2015. https://www.researchgate.net/publication/280696619_The_Internet_of_Things_for_Health_Care_A_Comprehensive_Survey. Accessed 6 July 2018.

U.S. Food & Drug Administration. Pediatric medical devices. 2018. https://www.fda.gov/MedicalDevices/ProductsandMedicalProcedures/ucm135104.htm. Accessed 1 Nov 2018.

Zhang P, Walker MA, White J, Schmidt DC, Lenz G. Metrics for assessing blockchain-based healthcare decentralized apps. IEEE. 2016. https://doi.org/10.1109/HeatlhCom.2017.8210842.

# Index

© Springer Nature Switzerland AG 2019
E. R. Onyejekwe et al. (eds.), *Portable Health Records in a Mobile Society*,
Health Informatics, https://doi.org/10.1007/978-3-030-19937-1

Printed by Printforce, the Netherlands